Landscript

地文

（韩）承孝相 ○ 著

黄晶涛　韩桂花 ○ 译

Landscript

The Inscription of Nature and Life on the Land

辽宁科学技术出版社
·沈阳·

图书在版编目（CIP）数据

地文 ／（韩）承孝相著；黄晶涛，韩桂花译．—沈阳： 辽宁科学技术出版社，2024.1
ISBN 978-7-5591-2498-2

Ⅰ．①地… Ⅱ．①承… ②黄… ③韩… Ⅲ．①建筑史－研究－西方国家 Ⅳ．① TU-091

中国版本图书馆 CIP 数据核字（2022）第 066525 号

出版发行：辽宁科学技术出版社
　　　　　（地址：沈阳市和平区十一纬路 25 号　邮编：110003）
印 刷 者：广东省博罗县园洲勤达印务有限公司
经 销 者：各地新华书店
幅面尺寸：170mm×240mm
印　　张：9.25
字　　数：150 千字
出版时间：2024 年 1 月第 1 版
印刷时间：2024 年 1 月第 1 次印刷
责任编辑：杜丙旭　周　洁
封面设计：周　洁
版式设计：周　洁
审　　校：尹　萍　陈明玉
责任校对：韩欣桐

书　　号：ISBN 978-7-5591-2498-2
定　　价：58.00 元

联系电话：024-23284360
邮购热线：024-23284502
http://www.lnkj.com.cn

**目
录**

1

2

未落成项目的记录　069
Document of Unrealized Projects

2002 年作为建筑师首次被韩国的国立现代美术馆选为"年度作家",并举办大型展览时,我提出的众多建筑语汇中有"地文"(landscript)一词,可在维基百科(Wikipedia)上搜索到 landscript 这个词,这个词是我创造的,也是至今我做建筑重要的方法论。我相信,对一个因场所而生的建筑来说,场所已经提供了所有的解决方法。

我出生在朝鲜战争中避难到釜山的家庭中。从 1950 年开始打了 3 年的朝鲜战争把韩国土地变成焦土,包括汉城(今首尔)在内的很多城市遭到破坏。之后韩国依靠西方国家的援助和军事独裁政权提出的强有力的经济开发政策,经济飞速增长,创造了世界经济前十位的奇迹。但是快速增长都是有代价的,贫富差距、阶级矛盾、环境污染很快跟随其后。特别是人们将近代化误以为是西欧化,盲目引进外来文化后摒弃掉自己固有的文化。其结果是韩国的老城里填满了急匆匆建造的杂乱无章的建筑,失去了传统风貌,有着悠久历史的大地被低俗的装饰所蹂躏。因此我们都不知道自己是谁,站在哪里,这就是 20 世纪后期的韩国人和城市的状况。这本书就是从对这种状况的省察出发的。

我是因在中国的第一个项目——长城脚下的公社,于 2000 年首次到访北京。当时的北京街头是电车和自行车覆盖的落后情景,但变成满街都是汽车只用了不到 5 年的时间。每次去北京我都能够目睹到,雨后春笋般起来的现代建筑下,过去的痕迹在渐渐消失。完成长城脚下的公社之后我一直努力通过我做的建筑交出不同的答卷,也许这就是我对磨灭记忆的开发所能做出的小小的抗拒。但是北京在通往着另一个世界,中国国内的其他城市也在天翻地覆地变化着,并且都在擦掉古老土地的纹理。看着这些现象我能做到的只有心痛,同时在 2009 年写下了这本书。

此书出版后已过了 12 年,最近在中国要重新出版,为此我有了新的感触。特别是最近在杨柳青大运河国家文化公园的大师邀请赛中以"地文"为主张提出的方案获胜,此番荣耀说明我长时间坚持的主张得到了认可,因此感到无比的成功和自豪,正好这时候邀我出这本书的中文版,我即刻答应下来。这又是无比的光荣。

土地并不是我们的消费品,活在这个时代的我们只是暂时借用我们后代该使用的土地而已,我们应该好好地使用,并记录下我们的生活,完好地交给下一代,这才是应有的道理。也只有这样,才能使我们的生活得到延续。

感谢为此书的出版付出努力的人们,特别是对黄晶涛教授工作室和辽宁科学技术出版社表示深深的谢意。

承孝相

2022 年 3 月 25 日

In 2002, I was selected by the National Museum of Modern and Contemporary Art of Korea as the first architect chosen for 'Artist of the Year'. During my solo exhibition at the time, I put forward the term 'Landscript' among many other architectural terms I asserted. One searching for Landscript will find Wikipedia describing it as a terminology that I created, and it indeed has become an important methodology of my architectural practice. I profoundly believe that an architecture is built with a conviction that the place it occupies over provides all solution.

I was born in a family that took refuge in Busan during the Korean War. The Korean War which lasted three years from 1950 devastated the Korean land and destroyed many cities including Seoul. Post war, Korea relied on Western assistance and strong economic development policies implemented by military dictatorship, achieving a steep and miraculous growth of being one of the top economic powers. However, superspeed development comes at a price. Gap between the rich and poor, gulf between the classes, and environmental pollution followed.

In particular, people mistook Westernisation for modernisation; foreign cultures were introduced and welcomed, while Korea's own culture abandoned. As a result, many old cities in Korea lost their landscape identity with improvised and promiscuous buildings, and the land of long history was covered with cheap decorations and trampled. This book starts from reflection and examination of Koreans and Korean cities in the late 20th century. We did not know who we are nor where we stand.

I visited Beijing for the first time in 2000 for my first project in China – Commune by the Great Wall. At that time, the streets in Beijing were filled with trams and bicycles, but it took less than five years to change to a road full of cars. Every time I visited Beijing, I witnessed modern architecture rushing in and out, the traces of the past gradually disappearing. Even after the Commune project, as a small resistance to the reckless and oblivious urban development, I have been trying to provide different architectural attitudes through my projects. However, Beijing was already far headed to another world and other cities in China were also changing dramatically, erasing the very texture of their ancient land. The phenomena was heart breaking and it led to the writing of this book in 2009.

It has been thirteen years since the book was published and now that the book is being republished in Chinese, my emotions are bound to be new. In particular, my proposal based on 'Landscript' recently won the competition of Yangliuqing National Grand Canal Culture Park in Tianjin and it feels like my long-standing proposition has finally been recognized and rewarded. Another incomparable accomplishment is to publish the Chinese version of this book, which I agreed with immense honour.

Land is not a consumer goods. Living in this era, we are all temporarily 'borrowing' the land that our future generations would use. We should make good use of it, record our lives, and hand it over to the next generation in good condition. I believe this is the right way, and the only way we can continue the life and memory of human.

I would like to thank those who have contributed in the publication of this book, especially Professor Huang Jingtao's Studio and editors at Liaoning Science and Technology Publishing House.

Seung H-Sang

March 25, 2022

自序

　　写这些文字是在去年年底，度过繁忙的一年时间之后有了几天的闲暇，正因为有人要我之前对城市和建筑演讲的内容，我自己也觉得有必要整理之前讲过的多少有些不同的内容。写正文其实都没有花三天的时间，但是为了这三天写出来的文字的出版，从开始到最后要我写这篇前言，竟然需要九个月的时间。中间并没有拖延出版，首先是因悦话堂出版社内部严谨的编辑原则，有不少内容被无情地筛掉而放慢；还有因为与负责英文的裴炯敏教授在审阅过程中进行多次争论，又花了很长时间；还有位美国人在校对裴教授的英文的过程中提出逻辑上的指点。大家都说现在已经不是我能随随便便出书的时候了。

　　是的。我其实至今只写了几本书，但对于韩国的建筑师来说算是出书较多的了。他们说我的这些书是无事杂谈，带着极端的偏见，有些是歪曲事实，甚至说是用不确定的内容下了虚构的结论。我只能接受这些事实。

　　我经常强调自己不是做学问的教授，而是做事的建筑师，由此来辩护自己的执念和先入之见。因为我心底里还是在想："所谓作家持有的主观不就是那样嘛。就算违背事理也应该被理解吧……"但是，请问，建筑师是作家吗？

　　当然世上有不少建筑师将建筑视为艺术行为，不断创作出艺术作品，远扬自己作家的名声。但是我从来没有说过自己设计的建筑是作品，因此从来没有认为自己是作家。组织好我们的生活方式是建筑设计的目的，如果把它看作自己的一个作品，就像放在相框里一样对待，那我们太轻视建筑中蕴含的那些珍贵的生活。我也没有胆量将建筑所在的大地看作一挥而就的画幅，因为害怕。或许在我设计的建筑里生活的人们，因选择了那块地，他们的生活和命运已与这块土地息息相关。我要做的事情原则上是在土地的需求之上增添他们的希望，因此我能做的就是花时间和精力去做这些事情，除此之外，我无能为力。写这本书就是为了说明这一切，但还是说这里有不少带偏见之处。不管怎样，通过这本书的出版，我又多方面地学到和悟到。

这些年，以"地文"为主题做过几个项目。是因为没有说服力吗？大部分都没有落地。有些是参加设计竞赛时明知对方不太容易接受，但还是主张这个主题，提交成果。并不是因为舍不得遗忘这些项目，只是觉得值得参考，所以在这本书中对这些项目进行了介绍，其实也担心成为配拙文的拙作。

这本书是自从《贫者之美》成为我的枷锁之后的第一本介绍建筑论或方法论的书。那么这些年我的建筑方法的逻辑达到了能够出版书的程度了吗？如果这样想的话，那是痴心妄想、自以为是、牵强附会。所以出这本书，从某种角度上来讲是我将把这不成熟的逻辑埋在自己的脚下，宣言自己总算站在新的起点上。我要重新起航。

感谢能够让我重新起航的我所爱的人们。

2009 年 10 月

承孝相

于履露斋

This text was written at the end of last year, during a few days of respite after what had been a very busy year. There had been many requests for material concerning my lectures on architecture and the city and I had myself felt the need to organize my thoughts into a book. In truth, it took less than four days to write the main text. But nine-months passed from the time I finished this four-day text to the moment I was requested to write this foreword. It was not that the publication had been delayed. First of all, following the principles of my publisher Yeolhwadang, its content went through a severe editorial process. I then had to spend an extended time debating with Professor Pai Hyungmin, who under very difficult circumstances, translated the main text. He had taken a very critical attitude in his translation of the book into English. Even my English editor relayed his harsh judgments about my logic. They all were saying that the time for me to write so easily had passed.

Indeed, though I had published just a few books, for a Korean architect, I had written a lot. For them, these books were inconsequential essays filled with unrepentant prejudices, sometimes skewing reality and even jumping to false conclusions based on uncertain facts. I can only but admit these judgments to be true.

I have often defended my prejudices and preconceptions by stating that I am a practicing architect not a scholarly professor. Weren't they the nature of the artist's subjectivity and disposition? I told myself shouldn't the artist be understood even if the logic was defective? However, I must now ask whether the architect is in fact an artist.

There are of course many architects who believe architecture to be an artistic practice, who consistently produce famous artistic works. However, I have never called my architectural design a work of art. Thus, I have never called myself an artist. I have spoken that architecture is the act of organizing our way of life. To speak of this as if it were a framed object of my own possession seemed to take the importance of this new life too lightly. Nor do I have the guts to draw in a stroke, as if the land were a picture, the architecture that will stand upon it. That is because I am afraid. Those who live in my buildings may have been destined to live in that way because of their fateful selection of this powerful site, and because it is my principle to weave their hope onto the requirements of the land. In this sense, besides physically bringing the time and man-power towards the work, there is really no subjective operation to talk of. I wrote this book to explain this, and yet I was accused of being immersed in my prejudices. Hence, through the publication of this essay, I have again realized many things and learnt a lot.

I have worked on several projects that claim the theme of "landscript." Were they unconvincing? Most of them have been unrealized. In a certain competition, I deliberately turned in a project that I knew would be difficult to accept, just to make my point. I included these projects in the latter part of the book not because there were too dear to my heart to throw away but because I believed them to be relevant examples of landscript. I must admit that I am anxious they will prove as poor a work as my writings.

This is my first theoretical, methodological statement since *Beauty of Poverty* became a kind of shackle to my work. Then, can I say that my architectural methods have acquired a logic that is worthy of a book? I admit this is delusionary, self-serving, and farcical. The publication of this book is in a way a declaration that I will bury this humble logic under my feet and thus stand on new ground. I shall start anew.

I thank those loved ones who have helped me to this new beginning.

October 2009

Seung H-Sang

at IROJE

永恒与不朽，是人类与栖居环境长久的纠缠。大多数人的寿命匆匆数十载；建筑的寿命从几十年到几百年不等，极少数不朽的作品能横亘千年；思潮的革新和技术的进步以百年为单位，近一个世纪以来更加频繁地推陈出新。那么，什么是永恒与不朽？是违背重力法则、高耸入云的超尺度建筑？是完美向心的几何形城市空间？是以效率和功能为导向、不分昼夜运转的城市机器？还是房地产逻辑下千城一面、"无问西东"的全球现代化进程？然而，迄今为止，上述大多数的尝试都以失败告终：我们赖以栖居的大地被粗暴地抹平、割裂、重塑，为了彰显阶层的等级，炫耀技术的"肌肉"，或是服从于某种看似势不可挡的经济"神话"——我们很难再感受到留在土地上的、祖辈生活的余温；我们狂热地改变环境，却不自信能留下存在的痕迹；我们对未来寄予厚望，却将一处处支离破碎、面目全非的栖居环境托付于后辈。

《地文》（*Landscript*）一书，正是敏锐地觉知到了人类社会对于环境的种种误判和改造，并尝试抛出问题的答案。在承先生看来，土地是永恒的，她属于历史、过去、未来，为所有人类所共享。千百年来人类生产生活印刻在土地上的纹理，也因此饱含了情感价值和伦理价值。承先生温情脉脉地写道："栖居是在土地上留下印迹，贮存回忆的过程……地文是一种生生不息的，一种要求外物不断附着的生命体。它有魂，有精神，甚至能言。"地文如同掌纹，是每片土地独一无二的，是长期以来人类与自然地域互动的结果，是特定生产生活方式的时空投射。建筑或许会推倒重建，片区或许会改造更新，但这既不是纯粹的"造新"，也不是简单的"修旧"，而是在新的社会经济和技术条件下，用场所连接场所，时间连接时间，通过对地文的传承和创新，让人类的生命记忆得以延续。

承先生出生于 20 世纪中叶的釜山，正是韩国城乡建设高歌猛进、西方文化快速入侵的时代。对于这片急速变化的大地，承先生与许多同时代的大师一样，以开放之心接受变革，以虔诚之心倾听历史，以赤子之心坚守原真。其实，在韩国发生的一切，又何尝不是中国的缩影。我与承先生相识于 2019 年，彼时我作为杨柳青大运河国家文化公园项目总规划师，邀请承先生参加大师工作营和竞赛。大运河是五千年中华文明留在国土上的重要地文，是南北文明互鉴与交融的纽带。2017 年，习近平总书记作出重要指示："大运河是祖先留给我们的宝贵遗产，是流动的文化，要统筹保护好、传承好、利用好。"这一指导思想贯穿大运河国家文化公园项目的始终。承先生联合团队提出的"杨柳青历史区再生计划"，通过对历史文化名镇核心保护区针灸式的"再活化"、对元宝岛根据地文"再组织"新的设施、对文化小镇根据水流痕迹"再开发"新的业态功能，最终"在土地上续写，我们生活的记录和故事"。

一千多年后，当我们再读《兰亭集序》——"后之视今，亦犹今之视昔……虽世殊事异，所以兴怀，其致一也"，当知人类与所栖居大地的情感连接，是不朽而常青的。我们有理由相信，对于地文的传续，将引领我们更加诗意地走进明天。

黄晶涛

2022 年 10 月

Eternity and immortality are words to describe the long-lasting relatinoship between humans and their environment. Most people have a fleeting lifespan of decades, the lifespan of buildings ranges from decades to hundreds of years, and very few immortal buildings can exist for a thousand years. Historically, ideological trends and technological progress leapt every hundred years, and innovations have happened more frequently in the past century. So, what is eternity and immortality? Does it mean a superscale building that violates the law of gravity and towers into the clouds? Does it mean a perfectly centripetal geometric urban space? Does it mean an efficiency- and function-oriented urban machine that works day and night? Or does it mean a global process of modernization that makes thousands of cities look alike? So far, however, most of the attempts mentioned above have ended in failure: the land we inhabit has been roughly flattened, cut apart, and reshaped, in order to demonstrate the hierarchy of men, to flaunt the "muscle" of technology or to submit to a seemingly unstoppable economic "myth". It is difficult for us to feel the warmth of our ancestors" lives left on the land. We are enthusiastic about changing the environment, but we are not confident that we can leave traces of our existence. We have high hopes for the future, but we entrust our fragmented and unrecognizable living environment to future generations.

The book "*Landscript*" keenly recognizes human society"s various misjudgments and transformations of the environment, and attempts to provide answers to the questions. In Mr. Seung H-Sang"s view, land is eternal, belonging to history, the past, and the future, shared by all mankind. The texture imprinted on the land by human production for thousands of years is also full of emotional and ethical values. Mr. Seung H-Sang affectionately wrote: "Dwelling is the process of leaving traces on a land and storing memories... Landscript is a living entity that requires constant attachment to external objects. It has a soul, a spirit, and even a voice." Landscript, like palm prints, is unique to each piece of land, the result of long-term interaction between humans and the natural lands, and the projection of a particular way of producing and living in time and space. Buildings may be demolished and rebuilt, and areas may be renovated and renewed, but it is neither purely "creating the new" nor simply "repairing the old." Instead, under new socio-economic and technological conditions, we seek to make connections between times and places, and through the inheritance and innovation of Landscript, the memory of human life can be continued.

Mr. Seung H-Sang was born in Busan in the mid-20th century, during the era of rapid urban and rural development in South Korea and the rapid invasion of Western culture. For this rapidly changing land, Mr. Seung H-Sang, like many contemporary masters, accepts change with an open mind, listens to history with a devout heart, and adheres to the origin with a pure spirit. In fact, everything that happened in South Korea is a microcosm of China. I met Mr. Seung H-Sang in 2019, when I was the chief planner of the Yangliuqing Grand Canal National Cultural Park project and invited him to participate in the master"s work camp and invitational competition. The Grand Canal is an important cultural heritage that has been preserved by Chinese civilization for five thousand years, and it is a link for mutual learning and integration between northern and southern cultures in China. In 2017, President Xi Jinping made an important directive: "The Grand Canal is a precious heritage left to us by our ancestors, a flowing culture, and should be well protected, inherited, and utilised in a coordinated manner." This guiding idea runs throughout the Grand Canal National Cultural Park project. The "Yangliuqing Historical Area Regeneration Plan" proposed by Mr. Seung H-Sang and his team, through the acupuncture-like "re-habilitation" of the core protection area of the historical town, the "re-fabrication" of new facilities based on the local Landscript of Yuanbao Island, and the "re-structuration" of new commercial functions in the town based on the traces of water, aims to "continue to write records and stories of our lives on the land".

More than a thousand years later, when we re-read the "Preface to the Orchid Pavilion Collection" — "Future generations would see today just as we view our predecessors… Although times and things have changed, the matters that trigger people"s emotions and their thoughts and interests are the same," we should realize that the emotional connection between people and the land they inhabit is immortal and evergreen. We have good reason to believe that the continuation of Landscript will lead us to a more poetic future.

October 2022

Huang Jingtao

Landscript

地文

1

伊卡洛斯的悲剧

    2001 年 9 月 11 日，发生了一件震惊世界的事件——纽约世界贸易中心被恐怖分子袭击倒塌（图 1）。在这起惊天动地的惨案中，近 3000 名工作人员瞬间丧生，并由此催生了伊拉克战争。在该战争中，有 4500 名美军和 9000 名伊拉克军丧生，还有 60 万平民百姓成为这场战争的受害者（"Iraq War Results & Staristics at December 10, 2008", US Liberal Politics, www.about.com）。然而，至今为止这场悲剧还未终止，还在威胁着地球村的和平。

    我并没有足够的社会和政治方面的见解能够讨论这场恐怖事件，但作

The Tragedy of Icarus

    On September 11, 2001, the World Trade Center towers in New York collapsed after coming under terrorist attack (Fig. 1). In an unprecedented and shocking disaster, nearly 3,000 people who came to work in this building suddenly lost their lives. It set off a train of events that led to the Iraq War, which has taken the lives of 4,500 American soldiers, 9,000 Iraq soldiers and 600,000 civilians ("Iraq War Results & Statistics at December 10, 2008," US Liberal Politics, www.about.com). The tragedy has yet to end as it continues to threaten the stability of our world.

    I do not have the sociopolitical expertise to debate this terrorist act. But as an architect, I had great interest in understanding the reason these towers were targeted by the terrorists. The obvious reason lay in its symbolic value. As the tallest structure in New York City, it was not only the symbol

为建筑师非常关注这些恐怖分子为何将此建筑作为攻击的目标。其原因必然是此建筑所具有的象征性的意义。这座城市的最高楼不仅象征着纽约，也象征着美国，更象征着资本主义体系和西方文化。恐怖分子认为击倒此楼将预示着西方世界的灭亡。但在这栋楼里开启当天日常工作的无辜的3000 名市民，仅仅因为其在最高楼工作，就失去了自己宝贵的生命。

城市中高耸的超高层建筑我们叫摩天大楼，英文叫 sky scraper，其意就是傲慢。什么样的自信能够去划破天际呢？是一种愿望吗？答案是肯定的，那种想对天空触手可及的意志是人类执着不渝的宿命。

人一出生就渴望行走，这种直立是区别于动物的关键所在，是违背重力法则的。与万物终将掉落在地的秩序背道而驰，渴望登高望远的意志既是人类历史的开端，也是文明的起点，记录下这一切的恰恰是人类技术的发展史。然而，为这些成就我们总要付出惨重的代价。

of the city but also of the United States, capitalism, and Western civilization. Therefore, the terrorists believed that by destroying this building they would send the message that the Western world had fallen. If so, the 3,000 innocent people who began an ordinary day in these towers lost their precious lives only because they were working in the tallest building in the city.

A building in the city that soars upward is called a skyscraper. The Korean term "macheollu" has the same arrogant connotation. What kind of confidence allows one to scrape the sky? Was it a wish? Yes, it was. The will to reach the end of the skies was cruel destiny.

Man wishes to walk from the moment of birth. This erect stance, which distinguishes him from other animals, is a denial of the natural law of gravity. The will to reach higher ground, standing against the universal order that all things must fall, was the beginning of human history and the commencement of civilization. The record of this will is itself the history of technology. Unfortunately, the price for these achievements has always been very high.

Such was the case of the myth of Icarus. Daedalus, his father, built the Knossos Palace on the

图1. 纽约世界贸易中心的悲剧。

由日裔美籍建筑师山崎实 (Minoru Yamasaki) 设计的这座 110 层的建筑，1973 年建成时成为世上最高的建筑，总建筑面积多达 82 万平方米，9·11 事件发生之前这栋楼和自由女神像是纽约当之无愧的象征。我心痛惋惜，那些恐怖分子的目标还不如是与人命无关的自由女神像。

Fig.1. The tragedy of the World Trade Center, New York.

This 110-story-high skyscraper, designed by the Japanese-American architect Minoru Yamasaki, instantly gained worldwide fame as the tallest building in the world upon its completion in 1973. In terms of space, its gross floor area spanned well over 820,000m². It was unquestionably the icon of New York—the Statue of Liberty being another—until it was destroyed by the terrorist attacks of September 11, 2001. Had the target of such terrorism been the Statue of Liberty it would not have involved so many casualties.

伊卡洛斯 (Icarus) 的神话便是这类代价的说明，相传伊卡洛斯的父亲代达罗斯 (Daedalus) 奉命建造爱琴海克里特岛的克诺索斯 (Knossos) 王宫，但代达罗斯却泄露了王宫里迷宫的秘密，愤怒的米诺斯 (Minos) 王将他们父子一起监禁起来。但聪明的代达罗斯试图拔掉飞进监狱来的鸟的羽毛做成翅膀，与儿子一起越狱。他告诉伊卡洛斯羽毛是用蜡粘的，飞得太高就会被太阳融化，粘贴的翅膀就会瓦解，因此嘱咐儿子不要飞得太高。但是，飞到监狱外的伊卡洛斯陶醉在完全从重力解放出来的兴奋中，忘掉了父亲的忠告，飞到了太阳近处。最终蜡化开，他掉进了大海里。这个神话讲的也许就是违背重力法则带来的惩罚。

当翻看《旧约圣经》里关于诺亚方舟的记录，在洪水时代之后畏惧生活在土地上的人们要兴建巴别塔（图 2）来躲避其他灾难，但是建塔的过程中因语言无法沟通，塔最终倒塌。为什么无法沟通了呢？这也是神对人

island of Crete. But when King Minos discovered that he had revealed the secret of the palace labyrinth he angrily threw Daedalus and Icarus in jail. The wise Daedalus planned an escape by creating wings out of the feathers of birds that flew into their jail. He warned Icarus that the feathers had been put together with wax and that if he were to fly too high the heat of the sun would melt and destroy the wings. Daedalus told Icarus not to fly too high. However, the moment Icarus flew out of jail, rapt in the freedom from gravity, he forgot his father's warning. The wax melted as he flew closer to the sun and Icarus fell to the sea. No doubt, this myth speaks of the consequences that followed those who disobeyed the law of gravity.

The pages of the Old Testament tell the story of the great flood of Noah. Here it is written that after the deluge the Tower of Babel (Fig. 2) was built by those who were afraid of living on the ground. However, in the midst of erecting the tower, it fell down because of problems in communication. Why couldn't people communicate with each other? Again, we may read this as a story of God's punishment of man's reckless longing to reach the heavens.

图 2. 彼得·勃鲁盖尔的巴别塔，1563。

犹太历史学家弗拉维奥·约瑟夫斯 (Flavius Josephus, 公元 37—100 年) 将它定义为"抵抗的建筑"，因为巴别塔标志着人类相信用自己的伎俩能够摆脱耶和华的愤怒。

Fig.2. Tower of Babel by Pieter Brueghel the Elder, 1563.

This structure, built upon the belief that by wisdom human can avoid the wrath of the Almighty, has been called an 'architecture of resistance' by the Jewish historian Flavius Josephus (37-100).

类想到达神的领域的冒犯给出的报复。

古代美索不达米亚地区等地建造的名为塔庙（Ziggurat, 图 3）的祭祀建筑非常高大，但其内部无任何空间，仅用石头和土来进行填满，由此可知这一构筑物只是以体现其高度为目的。以那个时代的技术和条件来看，为了建造这一构筑物需要投入大量的时间、财力和人力，但这一切只是为了追求其建筑的高度。

在那个技术并不发达的时代，想要建高层建筑物并非易事，如巴别塔容易倾倒，因此出现"塔"的建筑形式来满足对高度的需求。"塔"字源自印地语的窣堵波（stupa, 图 4），窣堵波原是保存佛陀舍利子的地方。作为佛陀圣骸安歇的地方，需要与其他普通建筑区别开来，因此要多层结构建高才行。但是在技术条件匮乏的时代，只能用普通建筑的缩小版——塔来替代。所以在韩国寺庙的中间建立的塔几乎是石塔，大体上是模仿多

The Great Ziggurat (Fig. 3) was a ritual monument built in ancient Mesopotamia. Its height and scale was immense but it had no interior space. Filled with stone and earth, the purpose of the structure was simply to rise up high. Considering the technological level of the period, a huge amount of time, resources, and man power had been invested. Yet the achievement of height was reward enough.

It was not easy to build tall structures without advanced technology. Like the Tower of Babel, tall buildings often collapsed. The tower hence emerged as an architectural form of compensation. The stupa has its origins in Hindu language. Its origins lie as the repository of the sarira of Buddha. As the resting place of Buddha's sacred remains, the stupa consisted of multiple stories, thus distinguishing it from ordinary buildings. But because of the limits of technology, the stupa was reduced in scale. Hence the stone stupas placed in the middle of temple grounds are thus architectural miniatures in the shape of multi-story wooden structures (Fig. 4). The stupa, seeking to secure its proper nobility, was designed to look like a tall building. Furthermore, following the rules of perspective, the stupa

图 3. 相传以人类历史上最初的城市而闻名的美索不达米亚地区乌尔的塔庙坍塌的现状 。

传统的宗教建筑分为 4 种概念：第一，像帕特农神庙或者城隍堂等"神居住的地方"（神殿）；第二，金字塔或者陵墓，纳骨堂等"死者的居所"（庙所）；第三，像犹太人的会堂等"聚会的场所"（会堂）；第四，最为普遍的"高处"概念等。属于墓址的金字塔，从概念上是与塔庙有区别的不同类型的建筑。

Fig.3. The remains of the Great Ziggurat of Ur, known to be the first city in history.

There are four types of religious architecture: 1) a temple, or house for Gods, such as the Parthenon and Seonangdang; 2) a burial chamber such as pyramid, mausoleum, and catacomb; 3) a chapel, a place for gathering like the Jewish synagogue; 4) a 'raised place' which is the most commonly found type. It should be noted that the pyramid, an Egyptian tomb, belongs to a different category from that of the Ziggurat.

图 4. 佛塔和宝塔 。

根据曼达拉现象建造的类似佛教寺院的佛塔原本是圆顶形状，后来根据地域风俗适度变形后出现多种形式，中国叫塔寺。印度的桑奇大塔——公元前 1 世纪，中国河南省嵩岳寺——公元 523 年，中国河北省开元寺——公元 1001 年。

Fig.4. Stupa and pagoda.

The stupa, built along Mandala principles, was originally a dome-shaped monument similar to Buddhist temples. Later, as it spread to other regions, it came under the influence of various cultures and customs, and saw modifications into diverse forms. In China it became known as pagoda. The great Stupa of Sanchi, India, 1st Century B.C., Songyue Pagoda, Henan, China, 523 A.D., and Kaiyuan monastery Pagoda, Hebei, China, 1001 A.D..

层木结构形式的小型建筑，实际上这些塔自以为是高层建筑，希望能得到合体的威严和地位。甚至这些塔根据不同的高度其大小比例进行变化，从透视的角度看显得更高。由此可知，不管在哪个地方或者哪个时代，对攀缘登高的渴望都是人类的本能。

was scaled to look taller than it actually was. Regardless of place and time, the longing to ascend was

a human instinct.

理想城市的梦

解决人类史上渴望登高的时代就是哥特时期。在建筑史上，哥特和罗马时期见证了最惊人的技术成就。在建筑方面，遇到的技术问题就是内部空间的营建，在这里最关键的任务是抵抗重力，将所营造的内部空间中最重要的屋顶支撑起来。建筑越古老，技术就越落后，其内部空间也越小，也无法开洞留窗户。为了建造比较特别的建筑，就要在高度上提高，这时墙体的厚度或许会比房间的大小还要大。解决困扰人类已久的这个问题的时代就是哥特时代，是真正称得上高技术的建筑时代。

哥特人搞清楚了重力的移位，研究出扶壁（buttress）和飞梁（flying

The Dream of the Ideal City

In the Gothic period, this pan-civilizational aspiration came to fruition. Along with ancient Rome, the Gothic witnessed the greatest technological achievements in architectural history. The first technological challenge of architecture was the creation of interior space. The crucial task was to lift the roof, the most important element in creating the interior space, against the pull of gravity. The more ancient the architecture, the less advanced the technology, and the smaller the interior and the openings. In order to build something special, to build it tall, the walls would sometimes become as thick as the room itself. Solving a problem that had frustrated architects for a millennium, the Gothic period proved itself to be an age of high-tech architecture.

Gothic architects understood the dynamics of structure and invented devices such as the buttress girder，图 5）这种特别的装置，并将它发展成为建筑的造型因素，打造出了非常生动和轻快的外观。穹顶（barrel vault）屋顶结构将沉重的屋顶荷载有效地传达到扶壁支撑下的柱子（pier）上，将重量位移至地面上。墙体已不再需要支撑屋顶了，只是用在区分室内外的界限功能上；窗户可以开大了，灿烂的阳光可以通过花窗玻璃照射进来，哥特式的空间使人类终于可以接近神的地位了。人类只要有时间和钱财，多高的建筑都可以建造，高度已经不是问题所在了，由此土地开始被忽视。

看似获得神性的哥特时代的人类想到达的下一阶段应该就是人类自尊的问题了，因此紧随其后的是文艺复兴时期。这个时期也被叫作人本主义时代，这个时代的人民并非我们现在对人民的理解。很明显，当时奴隶并不属于人类的范畴，只有极少数特权阶层享有人的身份和地位，因此，文艺复兴时期的建筑只属于少数特定的权力阶层。

and the flying girder (Fig. 5). These structural devices eventually evolved into formal architectural elements that created a dynamic and light exterior. The arch and the barrel vault effectively transfer the huge weight of the roof structure to the pier. Aided by the buttress, the pier then shifts the load to the inside and outside. The window can now be opened wide, bringing in the marvelous light of the stained glass. With Gothic space it was possible to approach the throne of God. With enough time and money, it was possible to build as high as one could wish. Height was no longer a problem. The land could now be ignored.

The next stage for Gothic man, which had seemingly attained deity, was self-reliance. Thus came the age of the Renaissance. In the Renaissance, sometimes called the age of humanism, the idea of man was different from our understanding of mankind. Obviously, slaves were not included in the category of man and only a few privileged classes were provided the status of man. Hence the architecture of the Renaissance was dedicated only to the powerful few.

Take for example, St. Peter's Basilica in the Vatican (Fig. 6). Its great axis begins from the

例如梵蒂冈的圣彼得大教堂（图6），轴线从柱廊（colonnade）围成椭圆形的广场开始，通过长长的教堂正厅 (nave)，再经过教堂外部走廊 (transept)，最后形成屋顶的交叉拱顶 (crossing vault)。此交叉点是这座建筑最重要之处，伯尼尼 (Gian Lorenzo Bernini, 1598—1680) 制作的漂亮华丽的铜制祭坛 (baldacchino) 就在此处上方。只有借助神之名的教皇才能占据这里，所以此建筑是为唯我独尊的教皇服务的道具。

文艺复兴时期成就了伟大的艺术作品，建筑方面的成就尤为惊人，其中帕拉弟奥（Andrea Palladio, 1508—1580）的圆厅别墅（Villa Rotunda, 图7）有着非常重要的意义。此住宅建筑可谓整合了西方建筑的历史，历经 400 多年，至今仍是西方建筑的核心教材。

在意大利维琴察的郊区小山坡上给隐退的牧师建造的圆厅别墅，将正方形几何平面用东西南北交叉的十字形通道进行分隔，正中间设置圆形空

colonnaded oval piazza and passes through its long nave. A crossing vault is created at the point where the axis meets the transept. This crossing is the crucial architectural point and it is here that Gian Lorenzo Bernini (1598-1680) placed his beautiful, spectacular baldacchino. With God as his guardian, the pope is the only person allowed to occupy this place. Architecture is thus an instrument that serves the pope, who would eventually be anointed the infallible man.

The artistic record of the Renaissance, the rebirth of the arts and the humanities in the West, was magnificent. Among architecture's great achievements, Andrea Palladio's (1508-1580) Villa Rotunda (Fig. 7) has particular significance. It is a house deemed to have integrated the whole history of Western architecture. After 400 years, it still remains one of its central texts.

Built for a retired clergyman on a small hill in the outskirts of Vicenza, the house has a simple structure. A dome is placed at the center of a geometric square plan divided by a cross-shaped corridor running north to south, and east to west. The central rotunda represented the master of the house. From this vantage point, he looked after the house, commanded the surrounding landscape,

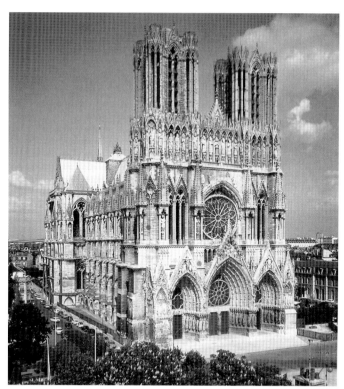

图 5. 哥特形式的兰斯大教堂（左），以及扶壁（右下）和剖面图（右上）。

支撑室内柱子和外墙的扶壁与飞梁相结合，起到柱子的作用，支撑巨大的荷载。通过拱形往外分散荷载，荷载消解后中间的空间变得自由，进而形成内部的拱廊，因此内部空间更深更丰富。

Fig.5. The buttress wall (right above) and section (right top) of the gothic Cathedral Notre-Dame de Reims (left).

The interior pillar and buttress (the latter seemingly supporting the exterior wall) are connected by the flying girder to act as a single structural element that sustains the massive weight of the roof. The use of arch frees certain parts from structural functions and the opening it creates is made into the interior arcade, adding depth and variety to the cathedral's space.

图 6. 梵蒂冈的圣彼得大教堂，1506 年开工，1626 年完工。

起初按照伯拉孟特 (Donato Bramante) 设计的希腊十字架形方案动工，后来的首席建筑师拉斐尔 (Raffaello Sanzio)、莎迦洛 (Giuliano da Sangallo) 等人进行局部修改，再次由米开朗琪罗 (Michelangelo Buonarroti) 和马代尔诺 (Carlo Maderno) 先后参与才完成。米开朗琪罗勉强继承首席建筑师的职位，但在那个时代建筑历史多少有些破坏的情况下，他重新认识到强调中心论的伯拉孟特方案的价值，并去进一步加强和深化。圣彼得广场的柱廊是 17 世纪中期巴洛克时代的作品，由伯尼尼 (Gian Lorenzo Bernini) 设计。

Fig.6. St. Peter's Basilica in Vatican City (Basilica Sancti Petri). Groundbreaking in 1506 and completed in 1626.

Although the foundation stone for this church was laid according to Donato Bramante's Greek Cross plan, modifications—including nave extension—were made by Raffaello Sanzio and Giuliano da Sangallo, and its final form was designed by Michelangelo Buonarroti and Carlo Maderno. Michelangelo, unwillingly taking the task as the chief architect of this monumental effort, nevertheless recognized the significance of Bramante's centralized plan and further developed its characteristics in his design. The colonnade of St. Peter's Piazza is a mid-seventeenth century Baroque addition by Gian Lorenzo Bernini.

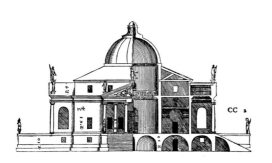
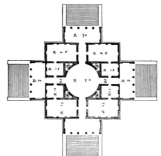

图 7. 圆厅别墅（1571 年完工，上）及剖面图（左下）和平面图（右下）。
根据居住者之名，也被称为卡普拉别墅 (Villa Capra)。这栋建筑是根据帕拉弟奥 (Andrea Palladio) 的
缜密的比例原则确定的，建筑本身即是周边风景重要的顶点。

Fig.7. Villa Rotunda (Completed in 1571, top), its section (above left) and plan(above right).
Also called Villa Capra after its owners, this residence by Andrea Palladio features a form governed by
the strict proportional principle set by the architect, and is itself the apex of its surrounding landscape.

间，结构如此单纯。圆形中心空间叫圆厅，是象征房屋主人的空间，他位于此空间，总管整个家，支配四周风景，创造了唯我独尊的体验。在这里大自然是要被征服的对象，建筑是支配的道具。不管基地如何，构筑完美的几何空间，确立无法与人分享的、世界上确信的自我，这种概念占据了文艺复兴以后西方建筑的主导位置。

这个概念又滋生出许多类似的建筑。比如到了近代，荷兰风格派 (De Stijl) 运动的代表成果也是代表着因内部空间的可变性引起空间内部结构的革命，里特菲尔德 (Gerrit Rietveld, 1888—1964) 的施罗德住宅 (Schröder House，图 8)，实际上是拥有中央空间的圆厅别墅的一种延续；革命性建筑师阿道夫·路斯 (Adolf Loos, 1870—1933) 试图利用内部空间的外部形式化设计出的很多房子也没能从中心性 (中心空间) 的意识中摆脱出来。

甚至 20 世纪最著名的建筑师勒·柯布西耶 (Le Corbusier, 1887—

and confirmed his stature at the center of the universe. Nature was an adversary to be controlled and architecture was an instrument toward this goal; whatever the form of the land, a perfect geometrical space was created. With the Renaissance, the concept of a consolidated self within a world that could not be shared with anyone else became the overarching idea of Western architecture.

This idea spawned a legion of architectural followers that extended into the modern period. Gerrit Rietveld(1888-1964)'s Schröder House (Fig. 8), for example, revolutionized the structure of interior space by introducing flexibility. It is a de facto extension of the Villa Rotunda and its centralized space. Even the houses of the revolutionary architect Adolf Loos (1870-1933), who sought the external formalization of interior space, could not escape the concept of centrality.

The same can be said of Le Corbusier(1887-1965)'s Villa Savoye (Fig. 9), the product of the greatest architect of the 20th century. Walking up and down the central ramp, enjoying the surrounding landscape, we experience the realization of Le Corbusier's architectural promenade. When one arrives at the roof garden, we discover that Le Corbusier has created the master's own new

1965）设计的萨伏伊别墅（Villa Savoye，图 9）亦是如此，柯布西耶在中央位置上设计的坡道是欣赏周边风景的建筑慢步路，可以走到屋顶庭院，里面构筑的空间是房屋主人自己的新世界，由此可知，20 世纪最高标准的住宅建筑也可看作是以帕拉弟奥式的现代方式完成的。难道是偶然吗？创造架空（piloti）的现代建筑浮在土地之上，将建筑和土地进行分开。

文艺复兴对中心论的迷恋并不局限于住宅建筑。比如，以威尼斯南部的帕尔马诺瓦（Palmanova，图 10）这座城市为例，中心论的概念尤其鲜明。认为周边都是敌人，为了阻挡敌人的入侵，在城市周围挖护城河，筑城墙。通过 3 道城门进城，会发现集中于一个中心点的辐射型路网。在几何中心，居住着身份地位最为高贵的人，这就是单一中心的城市。自然地，一个人越远离中心，阶层越低，身份越卑微。这样的城市完全是等级化和派别化的。

这种等级化的城市并不仅仅有帕尔马诺瓦。文艺复兴时期人们将这

world. I would underscore that this house, praised as the greatest work of 20th-century residential architecture, is the modern completion of the Palladian villa. This architectural masterpiece, employing the modern device of the piloti, floats above the ground. Was it mere coincidence? Henceforth, architecture was severed from the land.

The Renaissance obsession with centrality was not limited to residential architecture. For example, let us look at the city structure of Palmanova (Fig. 10) south of Venice. Here the idea of centrality is even more evident. Because its surroundings were all considered the enemy, a moat was dug out and a high castle wall was built around the city to ward off invasion. Entering the city through three city gates, the radial network of streets focuses on one central point. The highest ranking person lives in the center presiding over his domain. This is the so-called mono-centric city. As a matter of course, as one moves further from the center, the lower the class and the humbler the status. It is a thoroughly hierarchical and segregated city.

Palmanova was certainly not the only hierarchical city. Under the banner of utopia (Fig. 11),

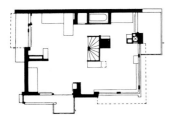 

图 8. 施罗德住宅（1924 年完工，上）和平面图（下）。

被联合国教科文组织列入世界文化遗产的此建筑，因立面形式成为荷兰风格派运动的代表作，但此建筑更加革命性的成就是，室内隔墙已不是支撑屋顶荷载的结构，因此可以灵活设置室内隔墙，能够打造自由的室内空间。

Fig.8. Schröder House (completed in 1924, top) and its plan(above).

This representative work of De Stijl, now a UNESCO World Heritage, is well-known for its elemantarized elevation which reflects the compositional principles of the Dutch art movement. Even more revolutionary, however, is its interior, where elements, freed from structural obligations, act as movable partitions to create flexible spaces.

图 9. 萨伏伊别墅（1931 年完工）的前面（下）和内部（左上）以及平面图（右上）。

此建筑如同教科书蕴含着勒·柯布西耶主张的"近代建筑五要素"： 1) 隔墙从结构柱子和承重功能中脱离出来，灵活的隔墙可以打造自由的内部空间和平面布局的原则。2) 为了使大地从建筑中解放出来，仅少部分与大地接触，这种架空形式最终达到整栋建筑被抬起来的效果。3) 外立面已不需支撑上面的楼板，因此能够任意组成自由立面的原则。4) 就像我们的两只眼睛在同一个水平面上，此建筑有着水平方向上展开的长条窗。5) 屋顶庭院可看作第二地面。

Fig.9. Villa Savoye (completed in 1931, above), its interior (top left) and plan (top right).

This work by Le Corbusier is a manual-like manifestation of what he termed "Five Points of a New Architecture": 1) the free plan, achieved through the separation of the load-bearing columns from the walls subdividing the space; 2) the pilotis elevating the mass off the ground, thus freeing the ground from buildings; 3) the free facade, as it is no longer related to the slabs or any other structural elements; 4) the long horizontal window, following the distribution of our vision; and 5) the roof garden, recognized as a second ground. The spatial experience of Le Corbusier's works is quite dramatic: various scenes and events abound in a lengthy promenade that encompasses both indoor and outdoor spaces, creating an impressive experience alternating between tension and release upon its visitors.

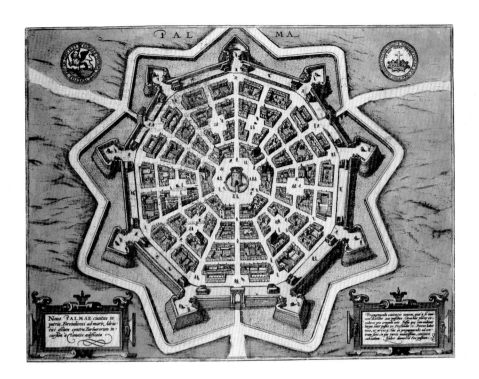

图 10. 帕尔马诺瓦的设计图。

1623 年斯卡莫齐设计的，为了防御威尼斯而建设的理想型军事城市。半径为 750 米，占地为 45 公顷的这座几何状城市，因顽固又无灵活性的结构，城市成长受限，最终导致衰退。

Fig.10. City plan of Palmanova.

Following Vincenzo Scamozzi's plan, this ideal military city was built in 1623 for the defense of Venice. However, its rigid, geometrical form—containing an area of 45 hectare within a diameter of 750 meters—proved to be inflexible to future growth and soon perished.

种城市冠以理想城市（utopia，图11）之名，在欧洲各处纷纷建设这种单一中心的城市。即便到了近代，城墙已拆除，当时的道路被林荫大道（boulevard）取代，但强调中心论结构的城市面貌如今依然可见。

理想城市是西方人观念的产物，它所反映的是根据脑中所想象的理想状态来实现的社会组织（图12）。为了将图变成现实，一块平坦的土地自然成为最佳的场地选择。因为自然被视为一个可征服的目标，因此崎岖不平的山地必须变得平坦，自然形成的水路不得不改为规则的几何图形，这样原始土地的野生状态就此消失，在被改造的土地之上强调的只有中心论、等级化和轴线，这便是完全的图解式城市。

在获得精神自由的法国市民革命和获得物质自由的英国产业革命下产生的近代城市，因人口暴涨需要新的城市秩序，由此迸发出无数的新未来型城市规划，但在这种近代城市中，等级性、中心性城市结构依然是重要

these monocentric cities were built all over Renaissance Europe. In modern times, the castle walls were taken down and large boulevards were built in their place. Yet to this day, the emphasis on centrality remains clearly visible.

The utopian city was the product of a Western conception. It was a diagram of a social organization created in the mind (Fig. 12), a diagram that was realized in material fashion. After the diagram was completed in the mind, the search for the site to build the city led appropriately to a flat piece of land. Because nature is a mere object to be conquered, an inconvenient topography would have to be leveled and a natural waterway would have to be turned into a geometrical composition. The savagery of the original land had to disappear. On this transformed site, only centrality, hierarchy, and axis were to be accentuated. This was a completely diagrammatic city.

The modern city was borne from the political revolution in France and the industrial revolution in England; the former brought psychological freedom and the latter material freedom. With the explosive growth of population in the modern city, a new urban order was required. Subsequently,

图 11.1516 年托马斯·莫尔创作的《乌托邦》里记载的题目为《乌托邦岛》的图片。

"utopia"的"u"来自希腊语，有两种发音，一个是表示"no"的"ou"发音，还有表示"well"的"eu"发音。由此可知，理想城市有着全然相反的意思，就是虽好但不存在的土地 (topos)。

Fig.11. A picture from Thomas More's Utopia, 1516.

The "u" from "utopia" comes from Greek, and is pronounced both "ou" (meaning "no") and "eu" ("well"). Hence utopia, ideal city, is an oxymoron—"well" but "non" existent topos.

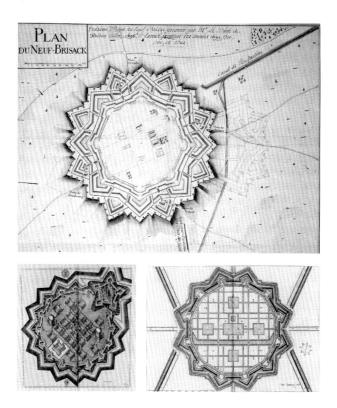

图 12. 文艺复兴时期的图解式理想城市方案。

16 世纪涌入的理想城市的规划，虽源自军事上的需求，但规划内容还是文艺复兴时期发展起来的透视法在城市中体现的结果。

Fig.12. Diagrammatic ideal city plans from the Renaissance.

Ideal cities suggested widely in 16th century were planned for the purpose of defense, and all of them aimed to completion of perspective view in city.

的概念。将城市分成市中心和次中心以及郊区，用各种颜色来区分居住区，商业区、工业区等比比皆是。提出科学的统计数据，在理性和合理之下画出的总体规划（图 13）也是相同于文艺复兴时期理想城市的蓝图。为了成就新理想城市的梦，提出的词是效率性和合理性，功能和速度。土地的伦理？在他们看来这个词是如此生疏。

　　将所有的假设用科学数据伪装的总体规划，到了 20 世纪，建设新城市时，成为全能的道具。无数的新城市都是根据蓝图建设的，他们相信梦中的未来理想城市终于可以落地了。但是不到半个世纪，这样的城市开始遭到批判。法国哲学家列斐伏尔（Henri Lefevre，1901—1991）曾猛烈痛斥他们，他说："在这种彻底图解化的柯布西耶式的居住机器里，不会有冒险、浪漫和心跳，只能让我们互相分离并远离对方。"实际上，在这样的城市中犯罪现象越来越多，贫富差距越来越悬殊，各阶层之间的矛盾越

there was a rush of new plans that presented new visions. In all these modern cities, however, the hierarchical, central urban structure remained the key concept: the division of the city into the center and peripheries, the distinction of residential, commercial and industrial areas by yellow, red, and blue. Reason and rationality were claimed by citing scientific statistical data. Yet just like the Renaissance utopian city, the drawing of this city, the so-called masterplan (Fig. 13), was also a diagram. In order to realize the dream of the ideal city, efficiency and rationality, function and speed were the key words of this drawing. The ethics of land? This was an unfamiliar concept.

The masterplan hid all its assumptions behind the veil of scientific data. It emerged as an omnipotent instrument in the new cities of the 20th century. With countless cities erected on the shining promise of blueprints, many believed that the dream of the ideal city would very soon be realized. However, after less than half a century, these cities came under attack. Among their harshest critics was the French philosopher Henri Lefevre (1901-1991). In the Corbusian "machine for living in," Lefevre saw "no adventure, thrill, or romance." People are turned inward and away from each

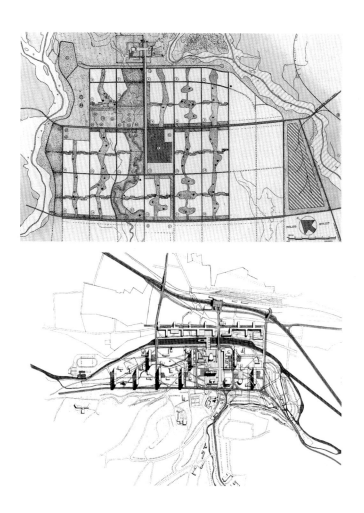

图 13. 勒·柯布西耶的城市规划方案。

以科学的统计数据为基础，以缓解城中心的混乱和高密度状态，快速的交通手段以及扩大植树面积等，作为重要的城市规划的原则。要建设如此理想的城市，就得找平地去做，然后将城市结构分成城中心和次中心，道路也从 V1 到 V7 分成等级，限定各级别道路的速度和宽度，主张连街道风景都有级别的等级化城市体系。

Fig.13. Urban plans by Le Corbusier.

Based on scientific, statistical analysis, his urban design was concerned with principles such as congestion relief and high-densification of city cores, smooth transportation, and enlarged green spaces. His ideal city stood on a vast flat land, hierarchically structured by the classification of city-center and sub-centers, roads categorized from V1 to V7 according to width and velocity, and stratified streetscapes.

来越深。

　　终于到了 1972 年 3 月 16 日，美国圣路易斯州政府爆破掉建成 17 年的普鲁蒂 - 艾戈（Pruitt-Igoe）住宅区（图 14），由此标志着新时代新生活的居住机器时代宣告终结，这是 20 世纪现代主义的终结。

other. Indeed, crime in the city increased; the gap between rich and poor widened; and the conflict between the classes deepened.

　　We are by now familiar with the demolition of the Pruitt-Igoe apartments on March 16, 1972 (Fig. 14). Seventeen years after the promise of a new age and a new life, this living machine was declared impotent. It was the final sentencing of a modernism born in the 20th century.

图 14. 炸毁的普鲁蒂 - 艾戈住宅区。

取第二次世界大战战争英雄名字的这片新住宅区，是遵循近代建筑的先导勒·柯布西耶和他组织形成的近代国际建筑师会议的纲领设计的。共有 33 栋 11 层楼的住宅区，在 1958 年完工后，曾在建筑论坛上被评为最佳高层住宅奖。但后来因这里千篇一律的空间、无功能的公共空间等，发生各种犯罪和种族纠纷现象，变成最恐怖的场所。结果于 1972 年 7 月 15 日下午 3 点 32 分（这是查尔斯·詹克斯有意表明的具体时间），州政府将这里爆破掉。此建筑的设计师是第二年 1973 年完工的纽约世界贸易中心的设计师——山崎实。

Fig.14. Pruitt-Igoe Housing being demolished.

This housing project named after a World War II hero follows the town planning principles set by CIAM (International Congress of Modern Architecture) and its core member, Le Corbusier. The novel idea of a residential town with 33 eleven-story apartment buildings was praised, upon its completion in 1958, as "the best high apartment" by the Architectural Forum, but this endlessly monotonous environment with non-functioning communal spaces soon degenerated into crime-infested, racially segregate neighborhood. As a result, at 3:32 p.m., July 15, 1972 (exact moment made famous by Charles Jencks), the buildings were demolished by the federal government. The architect of this project, Minoru Yamasaki, also designed the twin towers of the World Trade Center, completed in the following year.

城市和土地

　　问题在于这种疑似被西方已摒弃的总体规划开始践踏我们固有的土地。西方的总体规划至少对城市目标有定位，它依托于构建城市社区的悠久传统和针对社会组织形式本质的激烈讨论。但彻底被政治和资本权力相互勾结建造的我们的新城市（图 15），将这样的总体规划当成某种宝贵遗产而盲目模仿并进行改造。新城市被建在有丰富、独特历史的土地上，但这些却被完全忽略。不，更准确地说，这些历史印迹成了必须被抹除的羁绊。伴随着这种新城市的到来，过去的生活居住环境迅速被消失。如果那里曾经有一座山，就必须被夷平；如果那里有一片谷地，就必须被填平；

City and Land

The problem was that this masterplan, virtually discarded by the West, began to trample the peerless Korean landscape. In the West, the masterplan and the city at least shared a sense of purpose. The masterplan was borne from a long tradition of building urban communities and the fierce debates over the nature of societal organizations. In Korea, however, the new towns (Fig. 15) were built on the collusion of political power and capital. They were the product of land transfigured by the blind belief that the masterplan was some kind of treasured inheritance. New towns were built on ground that sustained a rich and unique history. They were totally ignored. These historical traces were inconveniences that had to be erased. Together with their architecture, the old living places quickly disappeared. If there was a mountain, it had to be leveled; if there was valley, it had to be

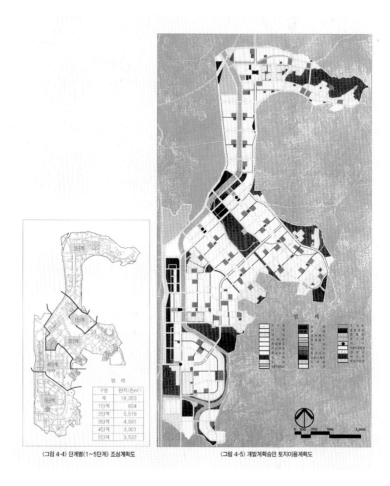

〈그림 4-4〉 단계별(1~5단계) 조성계획도　　　　〈그림 4-5〉 개발계획승인 토지이용계획도

图 15. 盆唐的总体规划。

这前所未有的规划，成就了四五十万人的新城市在四五年内像海市蜃楼般出现，这给其他新城市也带来巨大的影响，导致几乎所有的新城市规划都很相似，无法分辨。

Fig.15. Master Plan of Bundang, Seongnam-si, Gyeonggi-do, Korea.

The planning and construction of this new city with a population of 400,000-500,000 within a span of 4-5 years is historically unprecedented. Such hallucination-like invention nevertheless became an uncontested model for future new town developments, the result of which was the proliferation of undistinguishable cityscapes in these new cities.

如果那里有一条河流，就必须被改道。建筑师郑奇溶（1945—2011）曾经对此叹息道："曾经的肌理已消失，留下的只是开发面积。"但我们却认为这是新历史的创举而高兴。当然这里不存在任何社会共同体的概念，只是为了竞选时许诺的数字目标，这些目标是在他们任期内必须生产出来的课题，只有数量和数字才是重要的。这起初就不能被称之为一座城市，而仅仅是房地产的堆砌，是一座残缺的城市。

　　这样的城市应成败笔，并且仅一次足以。但是这种新城因房地产的升值让住户暂时获得幸福后，开始在全国范围内扩张，并且用相同的方法建造更多的城市，使人都无法辨别到底是哪里，这种没有社会共同体只有建筑群的怪现象占据并糟蹋着整片国土。

　　更大的问题不仅停留在新建城市上，而是在我们长久生活过的老城区也以再开发、城市整改的名义全都改成了千篇一律的模样。虽然我们的老

filled in; if there was a river, it had to be redirected. The Korean architect Chung Guyon (1945-2011) has charged that there is no longer any land, there is only area for development. Yet we rejoiced in our belief that a new era had been created. There was certainly no sense of community in the modern Korean city. Because the absolute goal was to reach a certain number — a politician's campaign promise that had to be delivered during the given term in office — only quantities and numerals were defined. This was never a city in the first place. It was merely an amassing of real estate, a crippled city.

Though this city should have failed, and though once was surely enough, the rise in real estate prices made its residents momentarily happy. These agglomerations sprang up all over Korea. Because all these cities were built in the exact same way, there was no way of distinguishing one place from another. Agglomerate monstrosities devastated the whole country. Here, there is no urban civility (civitas) only an accumulation of buildings (urbs).

The bigger problem was that this went beyond the building of new towns. In the name of redevelopment and betterment, old cities that had long been our homes were all razed. Even though our old cities sustained

城拥有的土地伦理和历史与西方人的图解式总体规划是截然不同的，但是现代韩国人认为建筑只是进行买卖的房地产，我们的老城和建筑应该是要被抛弃的旧恶而已。

"……首尔并非图解式城市，世界上有千万人口居住的城市大约有 20 座，但首尔几乎是其中唯一依山而建的，这也是首尔与其他超大城市——在平地而起的纽约、巴黎、伦敦，东方的北京和东京等从根本上不同的因素。"其实朝鲜王朝定都于汉阳（今首尔），至今记录着 600 多年的城市历史的首尔，其最大优势是依山。当时，郑道传和无学大师是看到汉阳有着内部围绕城市的四座山（内四山——北岳山、骆山、南山、仁王山）与外围的另外四座山（外四山——北汉山、龙马山、冠岳山、德阳山）相呼应的秀丽风光，就决定在此定都。在山势间蜿蜒流淌的汉江和支流的风景无比美丽。虽然现在的首尔已与朝鲜时代不同，众多建筑鳞次栉比，但山势的呼应和水渠依然证明首尔

a logic of the land that was different from the diagrammatic masterplan, modern Korea, which believed that architecture was merely real estate to be bought and sold, deemed them out-dated evils to be thrown away.

"... Seoul was based not on a diagram. Among the 20 cities in the world with a population of 10 million people, Seoul is the only metropolis located within a mountainous region. This is the element that distinguishes Seoul from all other metropolitan cities: New York, Paris, London and even Beijing and Tokyo in East Asia." Its mountainous geography is the most important element in recording the 600-year urban history of Seoul, a period beginning with its designation as the capital of the Joseon dynasty.

After the Joseon Dynasty was established, Jeong Do-jeon and the Great Monk Muhak decided on the location of the capital when they identified the magnificent landscape that surrounded Hanyang, the old name of Seoul. There are four inner mountains— Bugaksan, Naksan, Namsan, and Inwangsan—and four outer mountains— Bukhansan, Yongmasan, Gwanaksan, Deokyangsan. The Han River and its tributaries flow between these mountain ranges. It is a truly beautiful landscape. Unlike the Joseon Dynasty, Seoul is densely packed with many buildings. Nonetheless, with the

是世界上美丽的大都市之一。是的，首尔的地标是由山形成的秀丽风景。

"地标"这个词是1960年代美国城市规划理论家凯文·林奇（Kevin Lynch，1918—1984）在描述城市的时候使用的单词，这个词是西方人解说城市时使用的，不是用来描述我们的城市的。特别是在只能建设新城市的美国，在平地上建造是最容易的。所以这样的城市为了形成自己的特性，必须要有地标。因此，埃菲尔铁塔、大本钟（Big Ben）和摩天大楼等高耸的地标性建筑拔地而起，并主宰了这座城市。在东方，东京用庞大的城堡，北京用大尺度，确保城市的特性。到了最近也一直这样，在迪拜各种建筑不分青红皂白地在建设，在沙漠上用奇形怪状的人工建造物来打造城市印象，但我们的城市有着美丽山川，如果还像他们一样，那是很荒谬的。

我们的城市俨然不相同。美丽的山势已经成了首尔城市中的地标，建筑应该用适当的尺度来考虑不毁损这个地标——大自然。由小聚集形成的

harmony of the mountains and the water streams, Seoul is still one of the most beautiful cities in the world. The mountainous landscape is in itself the landmark of Seoul.

In the 1960s, Kevin Lynch (1918-1984) used the word "landmark" to explain the image of the city. It was a term used to explain the Western city not the Korean city. ... In America, where new towns were necessities, it was convenient to build on flat land. To create its own identity, it was necessary to build a landmark. Therefore towering monuments such as the Eiffel Tower, Big Ben, and skyscrapers rose up and dominated the city. In the East, the massive castles of Tokyo and the great scale of Beijing secured the identity of the city. It is no different today. The city of Dubai is being indiscriminately benchmarked by so many cities of the world. It is a place in the middle of the desert without any land pattern. Dubai is a city that has no choice but to create its image out of amazing man-made structures. It is ludicrous that out cities, with such beautiful mountains and streams, would seek to emulate this city.

Our cities are different. The beautiful mountains of Seoul are already the landmarks of the city. The old buildings of the city were built at a scale that would not damage the natural landmarks. The

集群之美才是我们应有的城市形象。如果我们对照朝鲜时期的地图和现代的地图就可以一目了然。19世纪画的《首善全图》（图16），像一幅山水画一样美丽。秀丽的山体，还有在山势之间流淌的水渠，在向阳处很多宅基地聚集在一起。从这幅地图上看，过去首尔的风光俨然是一幅画。

　　相反，在现代的首尔地图上看到的是，根据不同的土地用途用不同颜色分成几个地块，然后用红色将这些相连，这哪是人生活的地方？难道这不像是弱肉强食的战场吗？难道我的这种认知太过分吗？实际上按照这20世纪地图建造的首尔市现代面貌就是，各种地标奇丑无比，毁损山势，互不协调，相互对立。建筑为了凸显自己，表面做各种装饰导致自身都不协调，这样整体变成互相争分的情景，最后使首尔的风景不堪入目（承孝相，《如果我是首尔市长》，《月刊朝鲜》2008年8月号"附录"）。

image of our unique city is constituted by the collective beauty created out of the composite of small things. It is clearly visible when we compare the maps of the Joseon Dynasty with the maps of our modern times. The 19th century map of Seoul (Fig. 16) is itself a beautiful landscape painting. The magnificent mountains bring out their strength; streams flow between the mountains; homes cluster around the sunny slopes. The landscape of old Seoul was indeed a beautiful picture.

On the other hand, let us look at the modern map of Seoul. Areas are divided into functions by a mish mash of colors and connected by red lines. Is this a place for people to live in? Is this not a place where only the fittest survive? Am I being over-sensitive? In modern Seoul, a product of this 20th century map, all kinds of landmarks rise in gruesome fashion. They come in conflict with the mountainous forces, the true landmarks of the city, to create a disharmony. They clash among themselves to create a disharmony. They put on all sorts of make-up to make themselves more visible. Each is in itself a disharmony. And in this profane frenzy, the landscape of Seoul has become a jumbled mess." (Excerpted from "If I were the Mayor of Seoul," Monthly Chosun, August, 2008.)

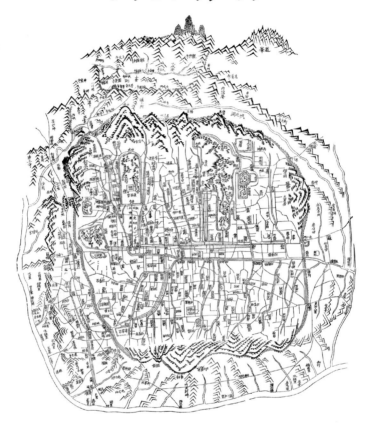

图 16.《首善全图》，1824—1834 年。
这张木版地图的名字取自《史记》里的"建首善自京师始"，简洁明了地表现出大自然中的秀丽的景色。

Fig.16. Suseon-jeondo. 1824-1834.
This woodcut map of Seoul accurately depicts through simple lines the beautiful features of the capital and its relation to nature. Its name derives from a phrase in *Shi Ji* (史记；Historical Records): "The construction of 'utmost virtue (Suseon)' comes from the capital (建首善自京师始)."

伦理的建筑

　　21 世纪刚开始的 2000 年，威尼斯国际建筑双年展的主题定为"少一些美学，多一些伦理"（Less Aesthetics, More Ethics，图 17）。我多少有些诧异，因为在西方建筑史上很久没有提到"伦理"这个词了。伦理是基于我和别人的关系的，在以自身的存在方式为主要目的的西方建筑中伦理肯定是生疏的词。但对我们祖先来说伦理是必需的法道，我们祖先做建筑的时候首先考虑土地和建筑之间的伦理，考虑建筑和建筑之间的伦理，考虑建筑和人之间存在的关系；我们祖先也不曾使用蕴含劳动之意的"建筑"这个词，而用"营造"来表达经营创造的意味；他们认为房子并不是物理性的砌筑，而是通过

The Architecture of Ethics

In 2000, the international architectural exhibition of the Venice Biennale announced its theme of "Less Aesthetics, More Ethics" (Fig. 17). I was a bit surprised. "Ethics" was a word in Western architecture that had long slipped my memory. Ethics emerges from the relation between the self and others. For Western architecture, dominated by methods of self-existence, ethics had become all too unfamiliar.

For our ancestors, ethics was a necessity. When our ancestors built, the first principle of architecture was to consider the ethics between land and building, between building and building, and between man and man. Thus our architecture was in accord with nature, with its surroundings, and with man. It created a landscape in which the whole was in harmony. At the beginning of

图 17. 第七届威尼斯国际建筑双年展的海报（右下）和被邀艺术家汉斯·霍莱因的作品（上）。

他的作品仿造日本龙安寺（左下）的寺院寂静空间，或许，他想表达的伦理就是东方建筑的概念。

Fig.17. The poster of the seventh Architectural Biennale, Venice (above right), which featured the work by Hans Hollein, invited artist (top).

His work was a parody of Ryoanji Temple's silent and atmospheric garden (above left). One can assume that architecture of ethics, at least for him, must have had a link to Asian values.

思维的过程建造的。这些都是我们建筑技术的要领，因此我们的建筑跟大自然协调，跟周边呼应，跟人和谐相处，演绎整体和谐的风景。相反，总在强调支配和服从的西方建筑终于在这新的时代开始寻找新的方向。

在西方文化中占据重要位置的文艺复兴，促使我们再用透视图的形式来谈论一番。透视图是文艺复兴时期的建筑师菲利波·布鲁内莱斯基（Filippo Brunelleschi，1377—1446）创造的画法（图18），在这张图上的所有事物和空间均汇聚于站在中间观察的人的眼睛里。换句话说，可以用无数角度表现的世界里，透视图上表现的是仅一个角度捕捉到的空间，透视图上叫SP点（standing point），在这一点上的人才能看到这个角度，离开这一点就不是画面上的空间，因此透视图画的是与别人无法共享的世界。拉斐尔（Raffaello Sanzio，1483—1520）画的《雅典学派》（图19）也同样，虽没有画出SP点，但这不朽的杰作也是基于透视图的画法，所

this new age, Western architecture, which had always emphasized dominance and obedience over harmony and accordance, was searching for a new paradigm.

The Renaissance stands at the central point of Western civilization. Let us talk about the Renaissance in terms of perspective (Fig. 18), a method of drawing invented by Filippo Brunelleschi (1377-1446). Objects and spaces in the perspective converge on the viewing eye of a person standing in the center. In other words, a world of infinite angles has been caught in one single line of sight. It is a drawing in which only the viewer placed at the standing point can see the whole. In other words, the perspective depicts a world that cannot be shared. In Raffaello's The School of Athens (Fig. 19), no one is drawn into the standing point. In this masterpiece, under the strict rules of perspective, all lines converge onto a viewing eye that cannot be seen.

A world that only one can see – this was the way the perspective saw the world. It was the subjective principle through which the Renaissance saw things, a sentiment that influenced the centralized worldview of the West. In Ways of Seeing, John Berger (1926-) castigated the distorted

图 18. 布鲁内莱斯基的透视法。
这张图中的空间其实有无数个角度，但这张图中所有事物都消失到一个点，只有站在 SP（standing point）点的一个人才能看到这个角度。换句话说，透视图表现的是与别人无法共享的世界。

Fig.18. Perspective representation by Filippo Brunelleschi.
Among the numerous viewpoint from which one can observe a space, this perspective painting chooses a single standing point, building a pictorial construction where its represented world view is non-sharable.

图 19. 拉斐尔的《雅典学院》，1510—1511 年。

图中虽然没有画出 SP 点上的人，但图中所有的线条都消失到这看不见的人眼睛里。

Fig.19. The School of Athens by Raffaello Sanzio. 1510-1511.

Although no person is depicted in the standing point, the whole picture is constructed from the viewpoint of an invisible person.

有线条都汇聚在看不见的人的眼睛的某个点。

仅有一人能看到的世界，这种看待世界的方法就是透视图，这也是文艺复兴时期人看事物的主观礼数，这也对西方人的中心论世界观影响颇大。作家及美术评论家约翰·伯杰（John Berger, 1926—）在他的著作《观看之道》（*Ways of Seeing*）中曾痛斥透视图这种产物是怎样让我们扭曲地看待世界的；甚至已夭折的美国大地艺术家罗伯特·史密森（Robert Smithson, 1938—1973）将此搬到作品中，斥责这歪曲的世界观。

那么除此之外还有别的画事物的方法吗？在这个问题上幸亏我们的祖先曾给过答案。从 19 世纪的民画《书架》（图 20）中可以看到，这书架的空间并没有归结到一个汇聚点上，而是书架的每格都有着自己的汇聚点，并且在每格里的事物本身也有各自不同的汇聚点。很惊奇，要知道画是画家经过思考表现出来的，那么画这个书架图的无名画家至少相信世间万物都有自己的中心。

view that the perspective had produced. Robert Smithson (1938-1973) brought the critique of this distortion into his own work.

Is there any other way of depicting things? Fortunately, our ancestors have already answered this question. Let us look at the 19th century genre of Korean folk painting called the Chaekgeori (Picture of stationary) (Fig. 20). The spaces of the bookshelves are not coordinated along one vanishing point; each shelf unit has its own vanishing point. Furthermore, the objects placed within each shelf unit do not adhere to the visual geometry of the unit. Each object has its own peculiar center. This is a wondrous thing. A drawing is created when the object depicted is processed by the creative subject. The unknown artist who created this bookshelf drawing must have believed that all things in the world had its own center.

To elaborate, this artist did not believe in a mono-centric world. He believed that all people, all things have their own center. If we truly believe in democracy, shouldn't we create our cities and buildings in this way? This is the city of plural democracy.

图 20. 民画《书架》，朝鲜后期。

共有 26 格的书架图，每格都有独立的中心点，放置的事物也都有不同的中心点，这样的多种视觉效果反倒像现代美术。

Fig.20. Chaekgeori (Picture of stationary), a Korean folk painting. 19th century.

A painting of bookshelves with twenty six compartments. Similarities between this Joseon period painting and modern art can be found in the multiplicity of viewpoints, represented in Chaekgeori by respective vanishing points of each compartment and its contained objects.

进而可知，这位画家不相信单一中心的世界，他相信我们生活的世界应该由我们所有人、所有事物作为中心。是的，如果我们信奉民主主义，我们生活的城市和建筑难道不应该是这样画吗？这就是多元化的民主主义城市。

总体规划下的西式现代城市不同于多元化的民主主义城市，在这里有城市主轴，有中央大道、中央公园、中心广场、中心商业区等，这样一座阶级化和分类掌控的城市从总体上没有脱离封建城市的范畴。那么与此不同的城市在哪里？其实我们对城市的思考一直没能从以西方城市为主的城市历史观中脱离出来，如果我们打开眼界，就会知道在这世界上由多种多样社会群体组成的城市还是存在的。其中代表性的有菲斯 (Fez) 和马拉喀什 (Marrakesh, 图 21) 等摩洛哥的城市。

这是相信真主之下人人平等的伊斯兰城市。从这座城市的航拍图上看，城市里"L"形或者"口"字形建筑像蜂窝一样布局在城市的各个角落，

The modern Western city of the masterplan is fundamentally different from the city of plural democracy. A city with a central axis, central park, central plaza, central business district, a city with words for class and taxonomy has in principle not departed from the feudalistic city. If so, where are the different cities? Our idea of the city has been too long dominated by the Eurocentric point of view. If we open our eyes we will see cities with a variety of social communities. Fez and Marrakesh (Fig. 21) in Morocco are such cities.

Islamic cities believe that all are equal under Allah. From the air, the square and L-shaped houses are organized like a beehive. Streets are like blood vessels that form a labyrinth. They organize the city. Though there are small differences in size, these houses have a structure in which each is equal to the other. In this city of collective houses, there are no central parks and central plazas - no place of centrality, no central axis, and no main street. There is no taxonomic distinction between commercial and residential areas. The roads are narrow and when they are wide they become public gathering places or open markets. A group of small houses and a public building come together to form a

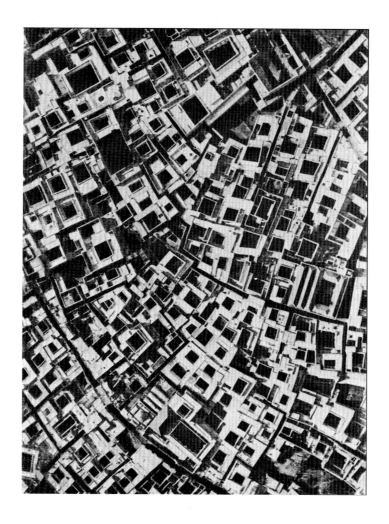

图 21. 马拉喀什的航拍图。

用西方的都市语言无法解释这座千年都市，所有个体都有独立的价值，因此其中一个也有同等于整体的价值。

Fig.21. Aerial picture of Marrakech.

In this millennium-old Moroccan city, its structure quite indescribable by urban terminologies of the West, all elements have individual value, and hence one and the whole have equal significance.

其间的街道如血管一般交错成了迷宫，它们共同构成了整座城市。尽管在规模上有细微的差别，但是所有建筑彼此在结构上都是平等的。在这座城市里，没有中央公园和中心广场等中心论下的场所；没有中心轴线，也没有主路，没有商业区和住宅区等区分；街道大部分狭窄，较为宽阔处成为公共集会地点或者露天市场；一系列小建筑和一座公共建筑聚合成一个最小的组团，然后这些组团根据土地标高的高低顺势而为，形成无数组群，进而再构成一座城市。因此，如果要看这样的城市不必转遍整座城市，看一个小组团便足矣。换句话说，你可以移除或者增加一个小组团，也无妨。这座城市在公元8世纪开始出现，我们必须强调的是，虽然1200年过去了，这些城市依然支撑着健康的城市社区体系。尽管建城时间悠久，但它们难道不是多元化的民主主义城市最真实的化身吗？

其实不仅仅菲斯或马拉喀什是这样的城市，只要建筑师或城市规划者

minimum collective unit. These units, adjusting to the levels of the land, coalesce to form the city. Therefore, one need not see the whole to see the city; one small unit may be enough to represent the whole. That is, one may take away one small unit or add another. These cities started to appear in 800 A.D. We must underscore that even now, after 1200 years, they continue to sustain healthy urban communities. Though built long ago, are they not the true embodiment of plural democracy?

Fez and Marrakesh are not the only such cities. Architecture without architects, cities without city planners are generally of such a nature. There is no need to go far to discover such a place in Korea. There are neighborhoods that we call "daldongne," the "hillside neighborhood" (Fig. 22). I often went to the Geumho-dong daldongne, which sadly no longer exists. For me, this neighborhood was a treasure trove of architecture. Though this community of economically disadvantaged people led a tough life, they knew how to live together. They created an architecture out of the wisdom of sharing. The sloped topography was a space in itself that produced all kinds of architecture. The dramatic layout of roads functioned not only as a circulation network but as a communal courtyard,

没有强行介入的城市大体上都是这种类型。也不用在别处寻找，我们周边也很多，像棚户村（图22）就是。如今已经消失不见的首尔金湖洞棚户区，是我曾经很喜欢去的地方，对我来说这里是建筑的宝库。虽然这里是贫困者的社区，但他们知道聚居的生存之道和分享的智慧，这些都如实地反映在建筑中。坡地的地形条件在空间上产生奇奇妙妙的建筑形式，有故事的道路不仅有着通行的目的，它又是公共交流的场所，可以根据时间和需求用于集会场所、游乐场、休息处等不同的功能区，所有空间都与居民的生活紧紧绑在一起。破旧又简朴的建筑并不是一次性建造的，而是体现着他们过去生活的记忆，可谓是活化石。谁说爱琴海圣托里尼岛的白色住宅区（图23）是天下最美的村落？对我来说，金湖洞棚户区在生活的原真性方面更加真挚而美丽。这是土地自身打造出来的建筑，我曾将此概括为《贫者之美》（图25），也以此作为我理解建筑的要点。

　　但是，过去遍布在首尔山坡上的各个小村庄，被再开发的伪善行为弄得支离破碎，变成了由混凝土堆流氓建筑占据的地方，形成极度不协调的情景。（图24）

a meeting house, a playground, and sometimes a resting house. They changed at will, functioning as a bond to everyday life. The ragged and humble architecture was not created overnight. It was a fossil through which the memory of the past could be imagined. Who called Santorini (Fig. 23) the most beautiful place under the heavens? The moon neighborhood of Geumhodong is more beautiful. It is an architecture molded from the land. I called it the "Beauty of Poverty" (Fig. 25), and it has been the basis of my architecture.

But these communities that once occupied the hillsides of Seoul were torn apart by the hypocrisy of our age called "redevelopment." Big bullies have taken over their place, buildings that are concrete piles of disharmony. (Fig. 24)

图 22. 现已消失不在的首尔金湖洞棚户区面貌。
并不在同一时间里建造的这些建筑可以推测过去。建筑是最单纯、简洁和简朴的。

Fig.22. Geumho-dong's hillside squatter neighborhood, Seoul, Korea, now destroyed.
Time could be trace from this gradually built townscape. Its architecture was most simple, modest, and succinct.

图 23. 希腊圣托里尼岛白色住宅区。

每个建筑既小又平凡，但下面一家房屋的屋顶成为上一家的露台，隔壁家的墙也是我家的墙，前面的小巷也是孩子们玩乐的院子。这座岛上村如此完好地体现共同体生活，它的空间结构与我们的棚户村极其相似。

Fig.23. White residential town of Santorini, Greece.

While the individual buildings are small and uncharacteristic, they together create an overall spatial structure which is a straightforward expression of their sense of community: the roof of a house becomes a terrace for the upper residence; houses share walls in between; alleys have a double function as children's playground. Such structure is quite similar to hillside neighborhoods in Korea.

图 24. 金湖洞再开发后出现的暴力状态。
最大容积率、最大覆盖率、最高层数、最小间距、最小停车数、最小公共设施等成为住宅楼的主要概念，在这里过去村子的痕迹是不值钱的伤感。

Fig.24. The brutal aftermath of Geumho-dong redevelopment.
Finding its raison d'être in profit-related values—such as maximum building ratio, maximum floor ratio, maximum height, minimum distance between buildings, minimum number of parking spaces, and minimum public facilities—these massive apartment towns leave no room for traces of the old town. A search for such is regarded as a symptom of cheap sentiments.

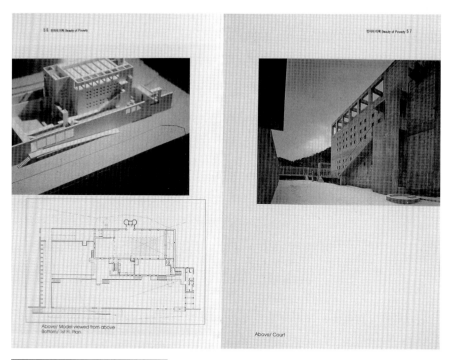

Above/ Model viewed from above
Bottom/ 1st Fl. Plan

Above/ Court

图 25.《贫者之美》（美建社，1996 年）。
1992 年 11 月由 14 位年轻建筑师组织的"4.3 集团"的展览——"这时代，我们的建筑"中，我提出这个主题并宣称这将是我日后的建筑理念。然后1996 年用同样的标题出版一本书，将我的建筑限定在这个理念中，然后我就释然了。

Fig.25. Beauty of Poverty (Migeonsa, 1996).
In the 1992 exhibition "Echoes of an Era," a collective effort by 4.3 Group consisting of fourteen young architects, I presented "Beauty of Poverty" as the central theme around which my works will be developed. Four years later, by publishing a short book of the same name, I further enhanced this self-imposed boundary... and liberated myself in it.

对土地的省察

对现代城市颇有关注的本雅明（Walter Benjamin，1892—1940）曾说过，自己喜欢读《尤利西斯》（*Ulysses*）是因为不管什么时候翻到哪一页都能洞悉整体的内容概要，包括他喜欢的城市也一样，在一座城市的任何角落都应该读懂这座城市。换句话说，局部不亚于全部，个体不亚于整体的城市，这种城市方能称为好的城市。

城市学者理查德·桑内特（Richard Sennett，1943—）在 1998 年密歇根大学进行的瓦伦贝格讲座（Wallenberg Lecture）中，以"民主主义的空间"（The Spaces of Democracy）为主题讲了如下内容：

Contemplating the Land

Walter Benjamin (1892-1940) was fascinated by the modern city. He said that the reason he enjoyed reading Ulysses was that whichever page he turned to he could understand the whole story. In the same way, he liked cities where the whole city could be known by looking at any of its corners. A good city is a city in which the part is not poorer than the sum and the element is no less essential than the whole.

In his 1998 Wallenberg Lecture at the University of Michigan, the urbanist scholar Richard Sennett (1943-) spoke of

> "The Spaces of Democracy." For Sennett, decentralized democracy does not aim to be a "unifying political force … Decentralized democracy is an attempt to make a political virtue out of this very

多元化的民主主义不以中央集权为目标，……彼此之间的差距是发展的主体。……多元化的民主主义亦有特别的物理形象，这种民主主义并不推崇庞大的建筑集群所表现出来的地标，而是追求在混杂的社会共同体中有多种语言层叠的建筑。……最终标榜多元化的民主主义的形象是彻底推翻体现整座城市的标志形象。……

　　换句话说，他想要的城市是从否定西方人过去用数千年时间打造的所有总体规划开始的。那么，如果建造城市的总体规划的手法有错的话，那还有什么正确的方法呢？

　　法国的城市学者弗朗索瓦·亚瑟（François Ascher，1946—）介绍过很有魅力的词。马塔波利斯（metapolis），这是相反于大都市（metropolis）的词。大都市作为近代城市的目标，是在功能和效率、区分和等级的基础上不断扩张的概念。他说如今以分享、沟通和共生作为时代语言的 21 世纪信息时代，

fragmentation …Decentralized democracy also has a visual dimension. This democratic vision may prefer the jumbled, polyglot architecture of neighborhoods to the symbolic statements made by big, central buildings … The result of visual, decentralized democracy should be, ultimately, to shatter those images which attempt to represent the city as a whole."

The city that Sennett wishes for begins with the denial of all masterplans that were created by the West.

The French urbanist François Ascher (1946-) introduced the fascinating concept of the metapolis as an alternative to the metropolis. The metropolis, the goal of the modern city, was a concept that assumed continuous expansion based on function, efficiency, and classification. Ascher argued that in the 21st century, in the new information age where sharing, communication, and co-existence have become the period concepts, the metropolis must be discarded. The metapolis, created through networking, must be the goal. I call the metapolis the contemplative city. It is a city of thought. But what should we be thinking about? We must contemplate the delusion that we own the

到了应摒弃大都市观念的时候了；他主张现代城市的目标应该是有连带关系的马塔波利斯，我翻译成"省察城市"。省察，省察城市，到底要省察什么呢？土地应是我们后代的所有，但我们误以为是我们自己的所有，然后忽视土地的伦理和它的生理状态，并将土地面貌仅用数据去评判，这些都是我们过去忽视土地伦理的肤浅行为。我想我们是否应该反省这种行为呢？

罗马广场（Römer Platz，图26）是罗马军队驻扎在法兰克福的历史场所，到了中世纪以后成为城市的中心，市政厅曾经在此地。当然这座城市也在第二次世界大战时因联合军队的攻击几乎成为废墟，罗马广场也被破坏得找不到痕迹，但是此处还是城市标志性的场所，是法兰克福市民们最希望尽早恢复的地方。他们起初想建造现代的购物中心，来体现经济的繁荣。但是后来市民们知道铝板表皮的商业建筑抹掉了这个历史场所的历史记忆之后，后悔当初建造这栋轻浮的建筑。

land when in fact it belongs to future generations; that we have ignored the innate logic and quality of the land; that we measured the land only by numbers. We must rethink our philistine attitude toward the ethics of the land.

Römer Platz (Fig. 26) in Frankfurt is a historical site where Roman legions were once stationed. After medieval times, it became an urban center with a city hall. During the Second World War, Römer Platz was totally destroyed by allied bombing. However, because of its symbolic value, it was the first place that the citizens of Frankfurt wanted to restore. At first, a modern shopping center was built to celebrate the city's economic revival. But after realizing that this aluminum commercial shell erased the memory of this historical place, they regretted their support for this superficial architecture.

Later on, a plan was devised to restore the buildings around the plaza to its historical origins. However, they again came to regret this plan to restore their proud past. The new "old" buildings turned back the city's history. Ironically, the restored buildings became obstacles to remembering all

后来他们重新规划和复原罗马广场周边的建筑，回到更加接近历史原貌的状态，回归他们荣耀的过去。但他们很快又后悔了，因为重新冒出的历史原貌并不能回顾过去罗马广场上发生的历史，反而退却了城市的历史，成为复制的场景而已，并不算是建筑。

后来到了 1980 年，在连接罗马广场和罗马大教堂的重要场所中规划文化综合设施，选用柏林出身的年轻建筑师团队 (Bangert, Jansen, Scholz, Schultes) 的设计方案，至此罗马广场迎来了新的局面。这座约 1 万平方米的综合性文化空间叫锡恩美术馆（Schirn Kunsthalle），由美术展览馆、音乐学校、美术工坊，还有几个住宿设施和小规模文化商业设施等功能区组成。他们的设计概念是，此建筑并不是单纯的一座建筑物或者标志性装置，而是连接场所与场所、时间与时间的城市有机体。从大教堂到市政厅在过去有道路之处重新恢复道路，过去有建筑的部分还是有建筑，继

that had happened in Römer Platz. This was not architecture but a frozen stage set.

Then in 1980, a new phase in the history of Römer Platz was initiated when a group of young architects from Berlin was selected to build a cultural complex linking the Römer Platz and the Römer Cathedral. The Schirn Kunsthalle was a 10,000 square meter cultural complex comprising an art museum, a music school, craft workshops, hotel facilities, and small shops. The architects Bangert, Jansen, Scholz, and Schultes approached this complex not as a single building or a monument but as an urban organism that linked place to place and time to time. They placed a road at the same spot where a road had formerly linked the cathedral and the city hall. They created appropriate linkages between inside and outside by placing buildings on the same site of previous buildings, and plazas where plazas had existed. Walking along the newly constructed street, one encountered the remnants of the Roman period, the remains of the Carolingian period, and the tragedy of the modern era. It was designed as an infinite travel through time in which one came face to face with the contemporary in reality and in imagination. The long and narrow colonnade was a sturdy 100 meter bridge that

续再恢复原来的广场，这样形成了室内外区分，再适当地将其进行连接。人们走在新铺就的道路，可以遇到罗马时代的遗迹，也能遇见加洛林王朝时代的遗迹，还能遇到近代的悲剧，这样现代的时间和痕迹能够在现实和想象之中进行碰撞，使人体验无限的时间旅程。有100多米长的又窄又长的列柱，就像在连接着大教堂和罗马广场之间暂时断掉的历史空白；在前面的院子里放置的过去时代破碎状态的废墟，看似随意废弃的状态，但在沉默中能够听到渗透在其中的过去岁月的故事。人们在这里观看历史层叠的状态，也能够很自然地感受到在历史展开过程中自己的模样。过去叫"Schwertvergäschen"（图27）的道路，如今在前面加上"过去的"（ehemalig）这个词来表示新道路的名字，因此这条路本身意味着一种历史记忆（参考承孝相，《建筑和记忆》《建筑，思维的符号》，石枕，2004）。这就是在此地拥有的历史之上介入新业态后的建造，是在找寻此场所过去的纹理后，再增添相关的新纹理。

crossed the brief historical schism that had separated the cathedral from Römer Platz. The pieces of old ruins, seemingly abandoned in the front court, told their old and silent stories. By actually seeing these layers of history, people could get a sense of their own identity within the flow of history. This street was called the Ehemalig Schwertvergäschen (Fig. 27). By placing ehemalig, "the former" in front of the street name, its memory was evoked. This is an architecture realized by placing a new program over the history of the land. It reveals the old patterns of the site, and in relation to this old pattern, adds a new pattern to the site. (Excerpted from "Architecture and Memory," in Architecture, the Sign of Thought, Dolbegae, 2004.) This architecture is realized by placing a new program over the history of the land. It reveals the old patterns of the site, and in relation to this old pattern, adds a new one.

图 26. 历史区罗马广场的恢复和再开发。
锡恩美术馆和附近遗址（左上），最初建设的购物中心（右上），罗马广场（左下），恢复成原来样式的广场周边建筑（右下）。

Fig.26. The restoration and redevelopment of the historical district of Römer Platz.
Schirn Kunsthalle and its adjacent excavation site (top left), the shopping center built at the very beginning of the redevelopment (top right), Römerplatz (above left), historically styled buildings surrounding the plaza (above right).

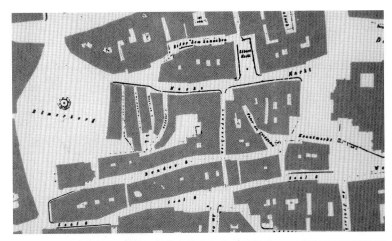

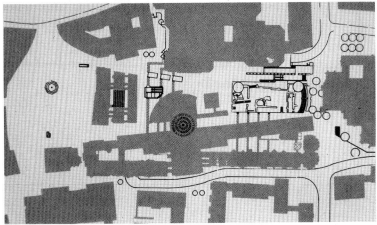

图 27. 第二次世界大战之前的罗马广场一带地图（上）和现在的地图（下）。
将两个地图相做比较，不难看出新建的锡恩美术馆和相邻新建筑群的布局，适当地改变了过去地图上出现的道路。

Fig.27. The maps of Römerplatz area: pre-World War II (top) and present (above).
A comparison between the two maps reveals that the Schirn Kunsthalle and its neighboring buildings have incorporated into their designs the town's old paths found in the pre-war map.

建筑和记忆

在韩语中，"teomuni"这个词是指印刻在土地上的纹理。"没有teomuni"意味着某些事情毫无缘由，真是有着惊人的含义。至少，这个词表明我们的祖先很清楚地认识生活与曾经一代代镌刻在这片土地上的历史不可分割，同时说明没有 teomuni 的生活就是无关于土地的生活。

无关于土地的生活是游牧式的生活，可能就是因为如此，我们的建筑和房子不再重视文化形式和家庭认同感的表达，长久以来已陷入房地产的买卖交易。我们只关注累积更多的财富，从一所公寓搬到另一所，过着游牧式生活。海德格尔（Martin Heidegger，1889—1976）宣称"人类通过栖居生存，

Architecture and Memory

In Korean, "teomuni" is literally a pattern inscribed on the ground. When one says that "There is no teomuni," it means that there is no basis or reason for something. It has a fascinating meaning. At the very least, the term signifies that our ancestors understood life as inseparable from the history inscribed onto the land.

Nomadic life is a life separated from the land. Could that be the reason? Our architecture and our houses, no longer values that express cultural form and the identity of the family, have long since degenerated into real estate to be bought and sold. That is why we live a nomadic life, moving from this apartment to that in order to amass more riches. Martin Heidegger (1889-1976) declared that being is achieved through dwelling and that only poetic man can dwell. Dwelling leaves traces on the

只有富于诗意的人才可栖居"。栖居是在土地上留下印迹，贮存回忆的过程。但是为了创造一段新的历史，印刻在我们土地上的回忆被视为陋习和恶行的产物，所以它们必须消失，我们就被迫患上健忘症，过着不关注土地纹理（teomuni）的、一种没有根基的生活。过往之所以存在，是因为我们经历过，但是我们认为这仅仅是一种匆匆路过。

阿多诺（Theodor W.Adorno, 1903—1969）在创造和使用"文化风景"（Kulturlandschaft）这个词时，对镌刻在土地上的历史之美道出如下解释：

"通过将景观概念与土地的严格关系中进行分离，用传统的城市性理解方法进行解释，才能够开启发现超越科学限制和参数存在的特性的可能……或许文化景观中蕴含的最为深刻的抵抗力在于被美化的历史，因为历史是经由往昔的真实痛苦铭刻而来……没有历史的记忆，就没有美丽而言。过去与之伴随的文化景观会清白地缔造我们的人性，让我们从宗派主义的束缚中解

land; it is a process that stores memory. Because the memory inscribed on our land was considered a product of bad customs and evil practices, they had to disappear. In order to create a new history, amnesia was forced upon us. We live a life without teomunee, a life without any basis. The past exists because we have passed through it. But we think that it has merely passed by.

Theodor W. Adorno (1903-1969) used the word "kulturlandschaft" to speak of the beauty of the history inscribed onto the land.

"By disengaging the notion of landscape from its strict association with land and interpreting it in a tradition of an understanding of urbanity may open the possibility to find qualities in the existing that lie beyond scientific limits and parameters … Perhaps the most profound force of resistance stored in the cultural landscape is the expression of history that is compelling, because it is etched by the real suffering of the past … Without historical remembrance there would be no beauty. The past, and with it the cultural landscape, would be accorded guiltlessly to a liberated humanity, free especially of nationalism."

脱出来。”

镌刻在土地上的历史……即是证明我们真实存在的景观。遗迹代表着历史流动的停顿，它是化石，是美丽的文化风景。杰克逊（John B.Jackson，1909—1996）在《废墟的必要性》（*The Necessity for Ruins*）一书中把传统景观的意义扩展到文化景观的现代理念中。“废墟既可以为复原提供线索也可以直接复原成原来的模样……旧秩序必须消失才能迎来景观的重生……这样历史将不复存在。”

1986 年在德国汉堡近郊的叫哈尔堡（Harburg）的地方，要在小广场上建造反法西斯纪念碑，艺术家戈尔茨（Jochen Gerz, 1940—）的作品被选中（图 28）。它是边长为 1 米，高度为 12 米的单纯的立方体，但惊人的设计概念是，它将每年以 2 米的速度沉到地下，最终会消失。纪念塔一般是为了记住某种事件而建立的永久构筑物，但在这里，塔是要消失的。哈尔堡的市民在这座

History inscribed on the ground; that is the landscape that makes us true. The ruin, a space where the flow of history has ceased, is a beautiful piece of fossilized landscape. In *The Necessity for Ruins*, John B. Jackson (1909-1996) expands the classical meaning of landscape towards the modern idea of cultural landscape. "That is what I mean when I refer to … the necessity of ruins to origins … The old order has to die before there can be a born-again landscape … History ceases to exist."

In 1986, the town of Harburg near Hamburg built a monument (Fig. 28) in a small plaza in memory of the town's resistance to the Nazis. The design by Jochen Gerz (1940-) was selected through a competition. It was a simple form, 1 meter by 1 meter square and 12 meters tall. Gerz's astounding idea was that every year, the pillar would sink 2 meters into the ground until it disappeared. We think of monuments as immortal obelisks that rise to memorialize an event. But this obelisk has now disappeared. As it sank, the citizens of Harburg would record on its surface the pain and oppression they had suffered under fascism. In 1992, exactly 6 years after its erection, the pillar was completely immersed into the ground, leaving only the memory of a monument that once stood

图 28. 哈尔堡纪念塔。

12 米高的塔逐渐下沉最终完全消失在地里的过程（从上至下）。

1983 年经过方案征集，约亨·戈尔茨和艾斯特·沙莱夫 - 戈尔茨夫妇的共同作品被选中。他们拒绝市政府要在公园中建此塔的建议，说服政府在哈尔堡市劳动阶层居住的杂乱街道的角落里建此塔。塔建成时有一个告示栏，邀请市民在这钢板塔身上做涂鸦，告示栏上这样写道："我们邀请哈尔堡市民和来这座城市的访客们在这里填写他们的名字，这样就能够警醒我们自己。在这 12 米的高塔上记录很多名字的同时，这座塔将逐渐往地里沉下去。有一天，这座塔将完全消失，抵抗法西斯的这座哈尔堡纪念塔的场地也将恢复清空。这意味着能够抵抗不义站起来的，只有我们自己。"

Fig.28. Harburg Mahnmal gegen Faschismus.

Serial photos of process of 12 meters high pillar sinking into the ground until disappearing(from top to bottom).

The husband-wife team of Jochen Gerz and Esther Shalev-Gerz was the winner for the 1983 design competition for the "Monument against Fascism." They argued against the city government's decision to erect the monument in a park, and instead chose a corner of a busy street in a working class district. The column, coated in lead, is accompanied by a signboard that reads: "We invite the citizens of Harburg, and visitors to the town, to add their names here next to ours. In doing so we commit ourselves to remain vigilant. As more and more names cover this 12-metre tall lead column, it will gradually be lowered into the ground. One day it will have disappeared completely, and the site of the Harburg Monument against Fascism will be empty. In the end it is only we ourselves who can stand up against injustice."

塔下沉期间，在塔身上用涂鸦的形式记录着受到的迫害和痛苦。所有悲伤的记忆被刻在塔身的同时，那些痛苦均从人们的身上脱离出来并被埋在地下。正好 6 年后的 1992 年，这座塔完全沉入地下并消失了，在那里只剩下这座塔曾经的记忆而已。由此换来了哈尔堡市民们的和解和原谅。

建筑师闵贤植（1946—）说过："曾经支配过我们整个社会的那些展示政治权力、宗教之力或者无形的资本权力的纪念碑式建造物，应在新的时代消失。"他也强调过："从土地拥有的条件中导出的形象成为因子后与周边形成统一的风景，注重人和自然的伦理，这种精神应该上升为，比起变化和整体更要讲究个体的身份和恢复日常点滴的新千年时代多中心多元化的时代价值，由此达到将天地人和大自然合为一体。现在的我们应要重新恢复我们的这种传统精神。"

摄影师盖尔斯特（Georg Gerster，1928—2019）出的书《空中摄影》（*Past*

there. The citizens of Harburg could now find reconciliation and forgiveness.

The architect Min Hyunsik (1946-) has claimed that "in the new era, monumental structures that symbolize political power, religious might, and the invisible force of capital must disappear … Our landscape must be integrated with its surroundings, created by figures pulled out from the conditions of the land. Such is an ethics of man and nature, a spirit that must be upheld as the goal of the new millennium—an era of multi-centered pluralism that places the identity of the singular, the restoration of the everyday over development and totality." Min has underscored that "we must today revive our old spirit of harmony with nature, a tradition that perceived the heavens, the land, and man as one."

The aerial photographs of ruins (Fig. 29) in Georg Gerster(1928-2019)'s *Past from Above* are moving testimonials. They are not just pictures. They make up a book that expresses the history imbedded deep in the land: it is a grand historical text, a monumental documentary, and a stirring saga. Buildings and cities, all irrevocably tied to the land, are patterns on the land, the communal *from Above*) 中记录着废墟的航拍图，非常震撼（图 29）。对我来说，那不仅仅是一本相册，它表达的是土地上印刻的历史，是壮大的史书，是雄伟的纪录片，是感人的小说。与土地有着密切关系的那些建筑和城市是在这土地上出现的纹理，也是在土地上书写的共同体的见证。

我们的废墟（图 30）给人的感动也不逊色。我们的旧城市和建筑跟西方有所不同，西方主要使用石材，所以残骸状态可以保留很长时间；我们的建筑使用土和木材，所以最终会完全消失，留下的只有几个奠基石和仅有的留白。但是这种留白的见证像哈尔堡纪念碑一样或者更加哀痛，因此久久留存。

日本的建筑学者香山寿夫（1937—）对场所记忆的重要性写了如下文字："场所可以用文化的积累及传统来认知，场所更应该是在不同的时间中进行不同事情的地方，是人类共同体统一合并的基础，场所是共同体的

testimonies of an era.

Ruins in Korea (Fig. 30) provide no less of an inspiration. Unlike the West, where the stone remains last for a long time, Korean old cities and buildings, made from earth and wood, almost completely disappear. With only a few foot stones remaining, Korean ruins present us with emptiness. For this reason, the testimony imbedded in this emptiness, like the lost pillar at Harburg, is more poignant and thus more lasting.

Hisao Koyama (1937-), the Japanese architectural scholar, has written on the importance of the memory of place: "Place can only be perceived as the layering of culture, that is, as tradition. In the continuous layering of time, a place is where many different events occur. It is the basis for a human community to come together as one. Place is the foundation and central support of community."

Let me bring in a few more words from Italo Calvino(1923-1985)'s *Le Città Invisibili*. "The city contains its past like the lines of a hand, written in the corners of the streets, the gratings of the windows, the banisters of the steps, the antennae of the lightning rods, the poles of the flags, every

基础也是支柱。"

　　再引用几段卡尔维诺（Italo Calvino，1923—1985）的《看不见的城市》
（*Le Città Invisibili*）的文字："城市不会泄露自己的过去，只会把它像手
纹一样藏起来，它被写在街巷的角落、窗格的护栏、楼梯的扶手、避雷的
天线和旗杆上，每一道印记都是抓挠、锯锉、刻凿、猛击留下的痕迹……
城市就在这种缜密的符号表皮中……在欲望之上继续赋予自身形态的城市
和被欲望擦掉或者擦掉欲望的城市……城市的形态的目录是无限的，所有
的形态都在找自身的城市直到新的城市继续诞生，所有形态变化完结时，
城市也开始终结。"对伊塔洛·卡尔维诺来说，城市是记忆和欲望互不可
分的交互交织的生物体。

　　是的，正如我们都有各自的指纹，往昔的记忆在所有的土地上都保有
印迹。有时它的纹路借由自然岁月而生，有时则是人类生命延续的雕琢，

segment marked in turn with scratches, indentations, scrolls … the city may really be, beneath this thick coating of signs … those that through the years and the changes continue to give their form to desires, and those in which desires either erase the city or are erased by it … The catalogue of forms is endless: until every shape has found its city, new cities will continue to be born. When the forms exhaust their variety and come apart, the end of cities begins." For Calvino, the city is a living thing in which memory and desire are inseparably intertwined.

　　Indeed, like our fingerprints and the lines in our palms, the memory of the past remains imprinted on all land. As every fingerprint is unique, so is the pattern of every piece of land. Sometimes its pattern is created from natural history; sometimes it is a pattern imprinted through the continuity of human life. The record and story of our lives are written on the land. The land is thus a grand and noble book of history, and thus is as precious as precious can be. Let us call this the landscript.

　　The landscript is an ever-changing organism and a life force that demands something be added

图 29. 乔治·盖尔斯特的《阿富汗的纳德阿里》，1977 年拍摄。

废墟，过去生活的记录。

Fig.29. The ruins at Nad-e Ali, Afghanistan. Photos by Georg Gerster, 1977.

Ruins—document of lives passed by.

图 30. 皇龙寺址废墟。

西方的废墟因由石头形成其残骸甚多，但我们的建筑用木材或者土来建造，因此很多时候建筑的痕迹都不好找。用尽生命后完全清空的场地，对我来说更加凄美。也许这就是建筑的目的，因为但凡看得见的就不是真相。

Fig.30. The ruins of Hwangryong-sa temple, Gyeongju, Korea.

Stone-built heritage of the West leave many remains, but the ruins of our traditional buildings, usually made of wood and earth, are hardly left with any trace. Exhausting their life in full, these buildings are now replaced by spaces of complete yet sublime emptiness. Perhaps this represents the ultimate purpose of architecture. What is visible is hardly the truth.

这是我们在这片土地上书写的生活的记录和故事。因此土地是一本壮丽而神圣的历史书，珍贵不已。这就是地文。

地文是一种生生不息的，一种要求外物不断附着的生命体。它有魂，有精神，甚至能言。据苏珊·朗格（Susanne K.Langer, 1895—1985）所说，如果建筑是一处场地性格的形象化的行为，那么建筑就必须要从认真倾听这一方经历过历史浸染的土地的新需求开始。建筑则是挖掘土地迷人语言的尊重性行为，这种行为经过深入的思考，然后以一种新形式的诗意语言谦逊地附着在土地之上。

建筑终将消失，纵使建筑被建造用来歌颂它的金主，在夸耀巨大技术成就之时，它们可能高耸入云，却只是为权势之人服务。然而任何人终有一死，建筑亦不可能打败时间。只有我们曾经存在的记忆能够留存，这便是唯一具体的真相（图 31）。

to it. It has a soul, a spirit, and it even speaks. According to Susanne K. Langer (1895-1985), if architecture is the visualization of the character of a place, architecture must begin by carefully listening to the new demands of the land. It is a land that has experienced the great expanse of history. Architecture is the respectful act of revealing its fascinating language; it is the act of thinking deeply and then humbly adding to the land by building on to it a new poetic language.

Architecture will inevitably fall. No matter how monumentally the architecture is built to glorify its patron for all time, no matter how high into the sky it may rise to show their power boasting a great technology, the architecture cannot defeat the law of gravity, as like everybody must meet their death. Only the memory that we were once there remains. That is the only specific truth (Fig. 31).

图 31. 2008 年受邀香港国际建筑双年展，以"地文"(Landscript)为主题参展的我的作品。在锈钢板上印刻我的建筑创作，再将钢板平放在地上。

Fig.31. My work, titled "Landscript," presented at the 2008 Hong Kong & Shenzhen Bi-City Biennale of Urbanism/Architecture. Corten steel panels featuring my works were laid on the floor.

Document
of Unrealized Projects

未落成项目的记录

2

Metapolis，大蒋谷住宅区规划，2003 年
与闵贤植、李钟昊、金荣俊共同合作

　　在这占地面积为 100 多万平方米，共有将近 2500 户的居住社区规划中，一开始我就想探寻不同于过去几十年的习惯性的设计方法。

　　这种项目一般都是由工程公司或者城市规划专业人士先编制总体规划，他们会先通过红色、绿色等颜色来区分土地利用性质，然后再控制建筑密度和路网等指标，编制总体规划后进行法定公示。这些手续走完之后，直接开始土建工程，削掉高坡，填补低洼地，砌筑挡土墙，并围好地块范围；景观设计也习惯用传统的做法，不会考虑因地制宜；路灯、桥梁等道路设施都直接根据标准施工图纸进行施工。之后建筑师们就会在地块上设计建筑，实施

Metapolis, Residential Town Plan, Daejang-gol,Seongnam, Gyeonggi-do, 2003
In Collaboration with Min Hyunsik, Yi Jongho, Kim Youngjun

Comprising 2,500 housing units in a land of 100 hectares, this project was an attempt to overcome the clichés of urban design that prevailed throughout Korea for the last few decades.

In building new towns it has been our custom to devise a so-called master plan, and this primary task was often placed in the hands of engineering companies or the so-called urban planners. The result took the form of land use plan—a colorful map representing zones for residence, commerce, industry, etc.—wherein its function-based divisions are designated with varying degrees of density. Adding a thread of roadways completes the plan, which was then put through legal procedures for public announcement. Civil works ensued, leveling the land and dividing it into sellable plots.

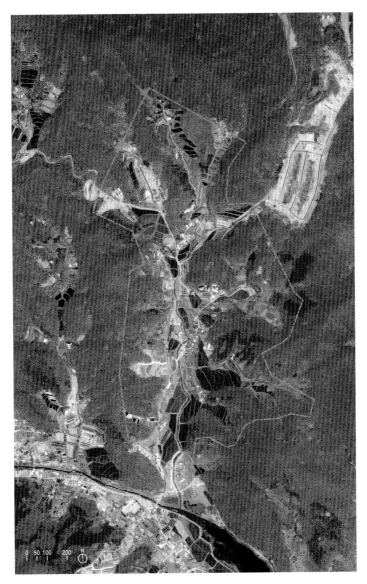

场地的航拍图。

Aerial photo of the site.

后建筑的外墙壁上就挂着杂乱的广告牌，这就是新项目从规划到实施的一个传统做法。

这种传统的做法中没有一个责任人从整体考虑整个项目，导致城市到处都是碎片化的风景，没有整体统一的城市风貌。这种现象貌似只有在这个国家能够看到，并且依然在持续，而且越来越多。由此一来我们行业的危机指数比任何时候还要高，行业处处都在怨声载道。但是，幸好还有人在这种大背景下继续探索着新的方式，追求梦想，才有了这个项目。

做设计时，首先思考的是这个基地所拥有的自身的特性，最有特点的是基地内保留下来的50多户农户。景观方面，基地周边被缓慢的山势所包围，在中间形成几处小的溪谷并向南开敞，小河在慢慢流淌，这样天然的地形自己已有独特的特点，并不需要再过度进行设计了。只要因地制宜，遵循这块土地自身拥有的逻辑，根据现有的50多户居民的居住形态，将居住人数扩

Customary landscape design had little room for the specificities of the site, as did the design of streetlights and bridges that relied on standard plans. The design of buildings on each plot was delegated to individual architects, but their façade was soon covered by a swarm of signboards.

This was the usual path through which a new city emerged. No one took responsibility for this routine process, as it was disjointed without a hint of integration. The city was but a combination of fragmented images. This crippled operation, a unique phenomenon of Korea, was constantly growing, and led to crises of communality, creating schisms in various parts of our society. In this context, it was quite a revelation to find a client who still sought a different approach to urban design.

My primary interest lied in the intimate stories of the land. The existing scenery, in which fifty or so farmhouses gathered to create a townscape of harmony, was mesmerizing. Surrounded by mountains of gentle slope and open towards the south, the site also maintained a charming landscape with its few small valleys and streamlets. The site did not require a landscape of spectacle. What was

方案模型的局部。

Partial views of the model.

至 2500 户即可。

因此首先在现状道路的基础上稍作调整，更新成为整个新居住社区的主要道路，再根据地形，塑造合适的居住形式并进行排布。整治现有溪谷的水渠，将水系两侧绿化廊道进行连接，形成三大系统（车行系统、绿化系统、水系）；在此基础上规划宜人步行道系统和将成为社区内主要交通方式的自行车道系统，就形成了五大完整的系统，即车行系统（red network）、步行系统（yellow network）、自行车道系统（silver network）、绿化系统（green network）和水系（blue network）。这五大系统在空间上进行交织，形成节点（node），并在这节点上规划有适当功能的公共服务设施。比如，在车行系统和步行系统交会的节点设置停车场或者咨询处等服务设施；在步行系统和自行车道系统交会处考虑自行车修理、自行车租赁或者学校等服务设施；在步行系统和绿化系统的交会处考虑布局书店或画廊等文化设施；在绿化系统和水系相遇之

needed was a simple multiplication of the number of its residences—in other words, an acceptance of the logic by which the old townscape was created.

This naturally led to the creation of road network that extended the logic of existing roads, and the formation and arrangement of housing types that responded to the natural geography. By adjusting the streams of the valley and connecting the green areas scattered throughout the foot of the mountain, three primary networks were generated, which was followed by the addition of paths for pedestrian and bicycle—the latter set as the main means of transportation. The crossing of these five networks—'red network' for vehicles, 'yellow network' for pedestrians, 'silver network' for bicycles, 'green network' for green areas, and 'blue network' for water streams—formed nodes, reserved for appropriate public facilities. For instance, parking spaces or information centers were placed at the intersection between vehicle roads and pedestrian ways, while bicycle shops or schools were sited at the meeting of pedestrian way and bicycle path. Intersections between green space and water streams provided sites for book shops and art galleries, while the crossing of green area and

处考虑生态体验场所等，这样在公共节点处布置公共服务设施会成为引发公共行为的场所，并均匀地布置在整个居住社区。

　　形成完整的系统后，开始考虑更多必要的设施。比如，为了维持我们对生命的尊重和虔诚度，在社区入口处规划一处骨灰堂；在居住社区的中心位置规划使用率较高的公共服务设施，并集中布置有老人的家庭；住房类型根据地形布置 7 种公共住宅和有多种可能性的独栋住宅，使得这里的居住类型更加多样化。

　　为了这个项目，找到了"rurban life"这个词。将"rural"（农村）和"urban"（城市）两个词进行合成，我想这个词能够代表这个新时代人们新的生活方式。在这前所未有通信网络信息高速发展的时代，日常办公都可以在线上进行，现在的城市反而更像是转变为周末的度假场所。项目所在地的区域也可能不是城郊住宅区，而是生产的产业区。当然，这些最后都不确定。这种有

waterways provided a natural site for an ecology park. Networks with diverse programs were joined to catalyze various communal activities, which were sheltered in public facilities evenly distributed across the whole town.

A combination of these systems formed the fabric of the city, and additional facilities were then inserted into the various parts of the town. For instance, a crematorium was placed at the entrance area of the town, functioning as an everyday reminder of the value of life. The center area was given much used facilities and units for the elderly, and to correspond to diverse geographical features of the site, seven different housing types, plus multiple single house districts, were presented, defying previous housing categories such as apartments, row houses, and single homes.

The program for this city could be represented by the term "rurban life". A mixture of "rural" and "urban," it refers to a new way of life already developing in this new age. With the advanced information transmission system everyday business tasks can be done online, and, as a result, our physical city is becoming more of a place for leisure and enjoyment. Hence this may be a town for

弹性的、不确定的但又一切皆有可能的城市——这种城市已不需要控制它整体规模——才是能够承载新生活的城市。我认为弗朗索瓦·亚瑟（Francois Ascher）的 Metapolis 项目就是这样的。

这个项目是否能实现，还未明确。但这项目的规划和设计理念在局部被公开之后，贪婪的开发商们一直聚焦关注此项目，导致基地的地价急剧上升，以至于这个现象最后涉及了国会层面，最终导致这个项目的前景变得更加模糊了。

industrial production, not a bed town. This is of course uncertain. This city of uncertainty, flexible enough for a broad range of possibilities—and thus no longer measurable through sheer physical size—is one that can contain the new way of life. This is what Francois Ascher called "Metapolis."

The future of this project is still uncertain. The moment the project was mentioned in the mass media there was an immediate rise of the region's land price, as greedy developers had been keeping an eye on this area for some time. None other than the National Assembly came to make an issue of the project's speculative potentials, and the project was put on hold.

新的地文（上）
和整体模型（下）。
New landscript (left)
and model (below).

光州亚洲文化殿堂，2005 年

其实这个题目本身就自相矛盾。文化和殿堂这两个本身就不是很搭的词，是因为要集中文化本身就有的专制和封建的属性，当然这与光州这座民主化运动的代表城市不尽相符。比如五一八望月洞国立公墓，是纪念对法西斯独裁抗拒而呼喊民主价值的标志物，但其设计真的很封建，由此可知，在我们的潜意识里的那些残存的思想是多么地根深蒂固。所以，这项目的开发商应该也是希望拥有庞大的具有标志性造型的建筑。

地理学家大卫・哈维（David Harvey，1935—）曾经说过："巨大理论并非是对总体真实的陈述。"也强调过："比起封闭而固定的理解方式，更

Asia Culture Complex, Gwangju, 2005

The very title of "Asia Culture Complex" is contradictory: the words "culture" and "hall" do not fit each other. This is because the idea of the centralization of culture contains a notion of autocracy and feudalism. Such idea also does not befit Gwangju, a symbolic area of Korea's democratization movement. However, even the 5.18 Mangwol-dong National Cemetery is a feudalistic design when it is supposed to portray the spirit of those who fought for democracy and against fascist dictatorship with all its might. This indicates that vestiges of feudalism are still deeply rooted in our subconscious minds, and it is probable that what the initiators of this project had in mind was an edifice of grand, symbolic form.

"Meta-theory is not a statement of total truth," remarked geographer David Harvey (1935-),

要追求开放的结局和具有辩证法的研究方式。"他也讲过，如今我们的社会比起形象（image）更重要的是叙事（narrative），比起美学（aesthetics）更重要的是伦理（ethics），比起存在（being）更重要的是生成过程（becoming）；最关键的是要做到"在差距中追求整合"。

要蜕变为"亚洲文化殿堂"的全罗南道厅地区是一个历史性的场所。全罗南道市政大楼的建筑本身就有保留价值，要求必须保留，但其实在这里要保留的不止是建筑本身。市政大楼前面的广场和周边的所有空间——小巷、小院、角落空间等都承载着悠久的历史，都是无比珍贵的。在这种空间里的建筑虽然破旧，但其丰富的形态和宜人的尺度与周边地区保持同质性，塑造了优美的风景。我认为这些所有的记忆都具有它的价值，我决定要保留这一切。

虽然因为要开发地下空间，地上原有的建筑物几乎都要进行拆除，但规

stressing the significance of a "dialectical research method and the pursuit of an open-ended rather than a closed and fixed method of understanding." He also noted that our society now places importance on "a plan that prefers narrative rather than image, ethics rather than aesthetics, and becoming rather than being," and that more than anything else we should now be able to "pursue integration despite our differences."

The surrounding area of the Jaollanam-do Provincial Office, to be transformed into the "Asia Culture Complex," has historical significance. The competition guideline stipulated that the Province Office, which is of some historical architectural value, should be left as it is, but this building was not the only thing that should be preserved. Along with the plaza in front of the Provincial Office, all spatial structures, such as allies, small yards, and corners were authentic and precious historical testaments. The buildings that form the spatial structures may be shabby, but the diverse volumes and friendly scales emit a sense of homogeneity with the surrounding areas, and as a whole create a beautiful cityscape. Memories contained in these spaces were all valuable and worth preserving.

划方案上新建的建筑计划着要在原来的位置上按照原来的大小进行重建。如果某处无法满足所需要的功能，就可以稍作变形或者与相邻建筑进行合并的方式来调整大小，但整体上外部空间还是尽量做到保留原来的尺度。

为了将整体紧密地连接在一起，设计了贯通全区域的媒体平台当作空中街道。这是各种信息传播的媒介，也是步行通道，因为媒体平台是新加的设施，所以从地坪上分离出来。

最终，原来在大地上的建筑和空间几乎都保留了下来，只是建筑的功能有了变化。在建筑中功能是次要的，场所决定了建筑，建筑功能只是载体而已。当然，这里的建筑功能随时都可以改变，重要的是场所以及基地的条件。

或许这种规划方案与项目的开发商所希望的"纪念碑式"有很大差距，也绝对不是他们想要的具有"象征性"意义的殿堂。但是我通过这次设计，成功保留了悠久和珍贵的历史文化，所以确信这个规划方案反而更有真实性，

Although almost all of the existing buildings had to be torn down due to underground development, our plan was to rebuild them on the same spot and in the same dimension. If a program required a bigger building some modifications could be made to individual buildings or a group of adjacent buildings could be skillfully combined, but the external space was preserved as much as possible.

In order to integrate all the areas of the site, a media deck in the form of an overpass was created that penetrated through the urban fabric. Used as a walkway and a medium to deliver various kinds of information, it was a new addition to what already existed and thus was logically separated from the ground.

Consequently, the fabric of the buildings and spaces related to the ground were mostly preserved, and what changed were their programs. Program is a secondary factor in architecture: it is elusive and subject to change at any time. Placeness of the site, the condition of the land, is what gives shape to architecture. Place determines architecture, and program is just there to aid the decision.

对光州这座城市的历史性空间带来的意义也更具有象征性。想必这个方案是针对他们的虚伪的想法做出的一个叛逆行为。

此方案在国际征集设计竞赛中获得第二名，所以让人遗憾。

This proposal was certainly not "monumental" in the traditional sense and may not have been "symbolic" enough to satisfy the powers that be. However, I was confident that this was the most monumental design, in terms of genuine commemoration, in the fact that through authenticity I was able to successfully preserve the city's long and valuable history. I also believed that this design brought the symbolic meaning of Gwangju's spaces to life, rebelling against past falsehood.

Unfortunately, we did not win this international competition, and the fact that we came in second place only added to our grief.

鸟瞰图。

Bird's eye view.

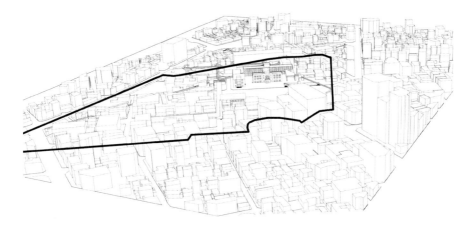

场地和周边的城市组织现状。
Urban fabric of existing condition.

场地和周边现状照片。
Existings condition.

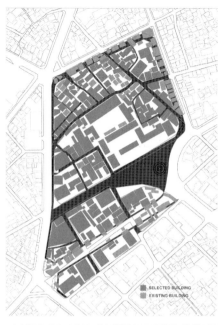

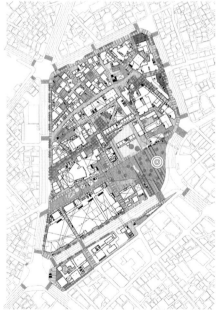

现状图（上）和上层标高平面图（下）。

Existing buildings (top) and floor plan of upper level (above).

Children's Museum

Asian Culture Center

Court of Knowledge

Visitor's Center

Asian Network Center

Artplex

Memorial Plaza

Sunken Plaza

Space Project

Asian Culture Creation Center

总平面。
Masterplan.

M-City，北京物流港城市居住区规划，2003 年

　　北京的主要道路系统由 6 个环道组成，其中的四环线和京津高速的交叉点，就是这个项目所在基地，原来是北京和天津港之间的物流区。随着开发条件的成熟，规划成为现代物流基地、居住区以及高尔夫球场等功能区。在整体 330 多公顷的占地中先做一块 33 公顷的基地，用来容纳 6000 多户的规划项目，邀请了扎哈·哈迪德（Zaha Hadid，1950—2016）、MVRDV 的威尼·马斯（Winy Maas，1959—）和我来参加设计竞赛。

　　加拿大的一家设计事务所做的总体规划整体很粗糙，没有明确的概念。他们彻底忽视这片土地上书写的历史，也许说他们无知更恰当。

M-City, Residential Zone Planning for Logistics Port, Beijing, 2003

Beijing's roadway system is constructed along a network of six ring roads, and the crossing between the fourth ring road and the Beijing-Tianjin highway is where the site for this housing district is located. Previously a distribution industry zone linking Tianjin port and Beijing, it was subsequently given a development plan that included a state-of-the-art distribution center, residential area, and a golf course. The design of a residential town for 6,000 units on an area of 33 hectares—one tenth of the whole site—was to be determined through an international competition among three invited architects/teams: Zaha Hadid, MVRDV, and myself.

The master plan provided by a Canadian firm was quite uninspiring—there was not a hint of a conscious concept whatsoever. It thoroughly neglected the history inscribed on the land, which

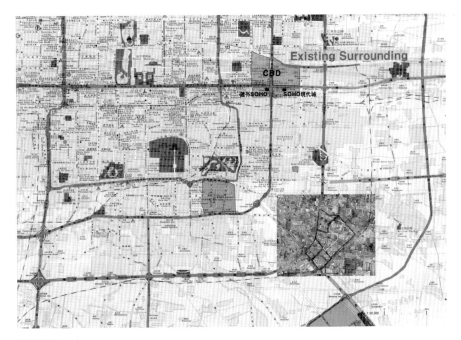

场地位置。

Site location in Beijing.

总体规划（上）和其中的本次规划范围
现状（下）。

Given masterplan (above) and existing
condition in overall masterplan (bottom).

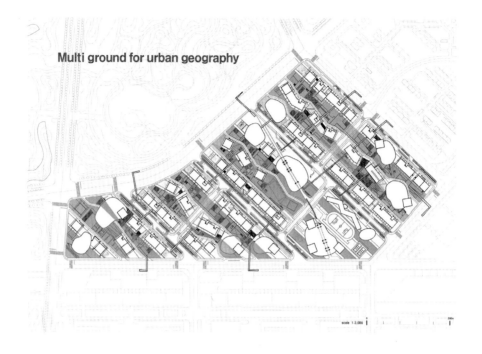

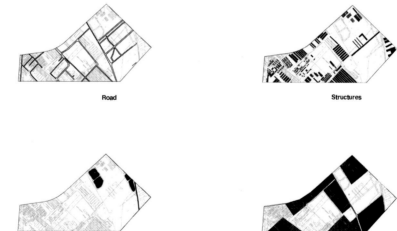

规划方案（上）和规划图层（下）。

Proposal (top) and layer of landscript (above).

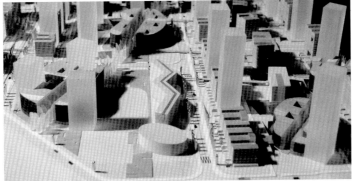

保留原有道路的规划模型。

Partial view of model to show existing road remained.

我从总体规划中去除了我所要设计的基地范围的相关内容，然后开始了解基地现状。基地内有现状水系，也有由茂密的杨树所围绕的长长的路，还有一直在耕种的农用地和虽破旧但温馨的民居。根据航拍图上的这些风景，我们参考航测图重新画出了地文信息，再把它分析成几种建筑语言，作为了后续规划的重要依据和指标。

　　为了连接南侧的文化设施集中区和北侧的高尔夫球场，在基地范围内均匀布置开放空间，将这些开放空间作为"脊椎线"（spines），居住空间则围绕着脊椎线进行排布。这些脊椎线属于公共空间，并在此空间中设置多样化的居民配套设施，由于规划的地面标高不尽相同，让人难以区分地下层和地上层，这种手法既可以避免每层可达性的差距，也可以打造多样化的场地空间，本身就形成了美丽的风景。

　　最重要的是在基地内保留原有场地记忆，由杨树所围绕的路是完全保留

seemed like a result of the designer's ignorance of rather than indifference to such issue.

I approached the work by discarding the previous master plan to uncover the actuality of the given site. In the land's aerial photo I found a long, poplar tree-filled road, along with streams, fields cultivated for many generations, and houses old, humble but attractive—existing elements that, with the aid of a detailed map, I could document as "landscripts" of the site, to be utilized as important design source and guide as they were analyzed and translated into architectural elements.

In order to form a connection between the southern region with dense cultural facilities and the golf course in the north, distributed across the site were rows of open spaces—which became "spines" encased by residential facilities. Spines, used as public areas with communal facilities for residents, were designed as multi-leveled platforms, not only to annul the distinction between ground level and basement—a distinction that leads to differences in accessibility—but also to create a beautiful and distinctive landscape with gradual rises and falls of the artificial ground.

The most important task was to preserve the memories of how life once happened on this land:

了；在现有居住群聚集之处设置胡同这一特殊要素；原有的水系和小巷也是以新的面貌重新出现在了原来的位置上。

此项目的名称定为"M-City"，同时找了很多"M"起头的单词：macro, mad, magic, magnet, major, manifold, manual, margin, max, master, mat, matrix, medi, meta, metropolis, micro, mid, mini, mix, mobile, modern, mono, moral, mosaic, motor, moving, multi……这些都是可以阐述这座城市的前缀。

但是，最终被选的是与过去的记忆和痕迹毫无关系的、用流线型公寓来填充空间的扎哈·哈迪德的方案。开发商劝我一同参与她的设计，但我做不到，不，是我不能做。

the road with poplar trees was thoroughly preserved; a special housing type called hutong(胡同) was applied to where a dense residential area once stood; waterways and small paths were reintroduced—albeit in a new form—in their original locations.

The title "M-City" has many connotations: macro, mad, magic, magnet, major, manifold, manual, margin, max, master, mat, matrix, medi, meta, metropolis, micro, mid, mini, mix, mobile, modern, mono, moral, mosaic, motor, moving, multi… all useful prefixes in describing this city.

However, it was Hadid who won the competition, her plan consisting of the usual warping forms that had absolutely no connection to the memories or traces of the region's past. The client of the project asked me to join as a co-designer in realizing Hadid's plan. It was an offer I could not—or rather should not—accept.

东大门国际设计广场，2007 年

　　东大门广场这片土地上书写的历史从首尔这座城市的角度上是非常重要的。这里是首尔城郭经过的场地，也曾是都城的边界，也是朝鲜时代训练都监的旧址，到了现代成了运动的摇篮，也是给国民带来诸多感动的地方。不仅如此，到了现代，随着周边的清溪川风景的变化，围绕此地产生了丰富的商业服务设施，由此成为了首尔最繁华和最有活力的场所之一。

　　但是如此巨大的运动场在城市中央本身毕竟是限制城中心功能的因素，因此更新改造是不错的建议，并且要改成具有设计内容的文化基地，在连接首尔的城市结构——奖忠坛文化地区以及大学路文化区——这一层面上也是

Dongdaemun World Design Park, Seoul, 2007

The history embedded in the area of Dongdaemun Stadium is very significant for the city of Seoul: it was the boundary of the capital city as it passes the old fortress of Seoul; it was the area where Hunryeondogam, the Office of Military Instruction of the Joseon Dynasty, was located; and it was the center of modern sports where memorable sporting events were held. Not only that, surrounded by many commercial facilities and nearby the recently transformed Cheonggye Stream area, it is still the most vibrant and busy region of Seoul.

However, the proposal to change the stadium area was appropriate because having a large-size stadium in the middle of the city restricts its function. The idea to transform the area into a culture zone for design deserves credit, in that it will integrate the city structure of Seoul by linking

该受欢迎的事情。尤其是此地区因杂乱的建筑和公共领域的缺乏已形成比较低级的城市风景，期待此项目能够给本地区带来健康而美丽的城市秩序。

但是最终被选的方案反而违背了这些所有的期待。建筑本身或许是漂亮的，但是这个方案与此地无任何关系。反而在相邻杂乱的商业建筑风景之上又增添了一栋"标志性"建筑，导致城市风景更加混乱。

我在此项目中将这片土地所拥有的3种记忆的恢复作为设计目标。第一是历史的复原，要重新建起古城郭，记起首尔600年的历史；第二是自然地形的复原，将场地东南侧的山坡通过建筑进行恢复，让人重新找回原有的场地记忆；第三是保留东大门运动场的记忆，局部按原来的结构重新建造。这样的方案是完全基于此场地上书写的历史重构起来的，历史本身就已经让此场所非常有意义。

当然，在这里建筑的形态是不存在的，只有记忆找回了形体，这源自对

Jangchoongdan and Daehangro culture zones. Even more, because this area maintained a rather poor urban landscape due to laissez-faire development of buildings and the lack of public spaces, the project—launched as an international competition among invited architects—provided an opportunity to reorganize the area in a more healthy and beautiful order.

However, the winning project was disappointing. The architecture itself may be beautiful but there was no connection whatsoever between the building and the meaning of the land. Rather, it enhances the complexity of the city scene as it added another "monumental" building to an already sprawled out commercial cityscape.

Our project aimed to revive three memories of the land: the 600-year-old history of Seoul (by restoring the old fortress); the original configuration of the ground (by using the building to recover the natural hill that once stood in the south-east side of the site); and the memories of Dongdaemun Stadium (by partially rebuilding its structure). It was a plan that superimposed all the history written on the land, itself becoming another important landscript.

这片珍贵的土地的热爱。东大门、清溪川、城郭以及运动场的记忆，除此之外在这里还需要什么更加标志性的意义呢？

Of course, the form of architecture, as always, cannot exist. Here, memory is the only thing with form, which originates from our affection and thoughts on this precious land. What other landmarks are needed when we have the memories of Dongdaemun, Cheonggye Stream, the fortress, and the Stadium?

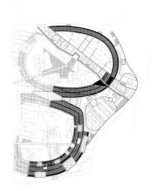

东大门运动场记忆的保留。

Reservation of memory of Dongdaemun Stadium.

原有地形的复原。

Restoration of original topography.

历史的复原。

Restoration of history.

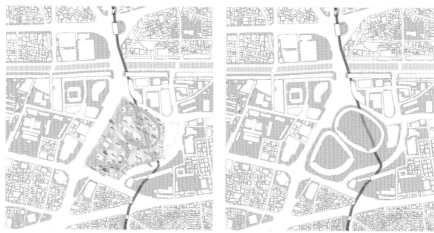

城市组织（上）和部分模型（下）。
Urban structure (top) and partial view of model (above).

效果图（上）和夜景（下）。

Perspective (top) and nightview (above).

献仁村，2008 年
与赵成龙、郑奇溶、闵贤植、李钟昊、金荣俊共同合作

　　此项目是以邀标形式进行的，是关于容纳 300 户居民的居住区的设计。位于首尔的南部献仁陵附近，原有住户聚集形成了村落，基地是北向的，但是在缓坡上，周边风景也很秀丽，作为居住区有着得天独厚的环境条件。建造时间并不长的原有住区，之前并没有任何城市规划的相关行为的介入，而是居民自己建造，最后形成的村子。其实这样的村落结构比任何一个城市规划师或建筑师建造的村落更加自然并富有丰富的空间，这并不是一件惊奇之事。在访问此地、勘察村落时，我很快就确定了这一点。沿着地形的坡度自然摆放房子，房子中间穿插着各式街巷和院落等，就像是游走在中世纪欧洲

Heonin Town, 2008
In Collaboration with Joh Sungyong, Chung Guyon, Min Hyunsik, Yi Jongho, Kim Youngjun

A closed competition was held for the design of this 300-household residential complex, located south of Seoul near Heonilleung. The site is facing north, but its gentle slope and beautiful scenery provide a very favorable environment for a residence town. There existed a small village for furniture manufacturers, and, although it did not have a long history, it was built by the residents themselves and not an outcome of a city development plan. It is not so surprising that this type of village feels more natural and is spatially more interesting than any towns planned by a city planner or an architect, and I was able to confirm this the moment I visited the village. The houses were naturally built along the slope of the land and in between them were marvelous allies and yards, and it felt as if I was looking at a small medieval mountain town of Europe. Inspired, I wanted to preserve the

的山中村里，我感动于此。因此，想积极保留这样的空间结构。

为了容纳300户规模的居民，只能拆除原有建筑，重新在基地内做规划。但是现状横穿整个小区的道路和其他几条道还可以作为后期住宅小区的道路使用，这样一来，原有道路的历史痕迹就可以保留下来了，并且这也是高效利用现状地形的一种方式。道路的保留是能够延续一个社区记忆的最好的方法，也是作为公共空间能够体现城市公共价值的重要的因素。

针对300户居住空间的设计，我想设计出多样化的模式，同时也想在差异中找寻协调的一种方式，并且觉得这也是对居住者的尊重。因此我邀请了5位建筑师，赵成龙、郑奇溶、闵贤植、李钟昊、金荣俊等与我投合的几位设计师分别负责5个地块，我则负责整体的协调工作，同时进行社区中央路以及商业等公共服务设施的设计。下一步的设计工作就是按照设计内容做出建筑体块，再把它结合空间组织设计进行布置，将5位设计师的设计方案布

spatial structure as much as I could.

However, because the existing village was incapable of holding 300 household units, the old buildings had to be inevitably torn down. But several streets including the street that crossed the whole site could still be used as streets for the new residential complex. As such, traces of the streets could be preserved and doing so was also the most efficient way to utilize the configuration of the ground. Streets, as a space for the public, are the most important factor in realizing the common value of a city, and its preservation is the best way to revive the memory of a community.

All 300 units were designed differently. By doing so, I wanted to pursue harmonization amid differences and respect the diverse lives of the residents. That was why I asked five architects to help me. I invited Joh Sung-yong, Chung Guyon, Min Hyun-sik, Yi Jong-ho, and Kim Young-joon, people that I had often worked with, to take charge of designing five different sections of the residential complex, and I was to coordinate the overall project while designing the center road and public facilities including stores. The design was finished by adding the volume from the program

置在规划总图上，然后再进行贯穿整体的景观设计，这个项目就顺利完成了。

但是在与建筑公司内部人员审查项目的会议上跟他们交流几句之后，我便发现我们的方案离他们的所愿距离甚远。他们想要中世纪欧风形式的暴发户式的建筑形态，而我们主张的保留每户零碎生活的记忆碎片和历史要素，在他们看来是没必要保留的，他们更希望将这片土地伪装成白纸状态再进行设计。

to the existing spatial structure, and the project was smoothly completed by placing the overall landscape architecture on top of the organized blueprints.

But in the jury meeting where internal members of the construction company reviewed the projects, even before sharing a few words with them, I realized that we were not on the same page. What they wanted was a posh architecture in an exotic medieval European style, and from this viewpoint, the land had to be covered up as a tabula rasa, and landscript of the old village's petty lives was something to be totally wiped out.

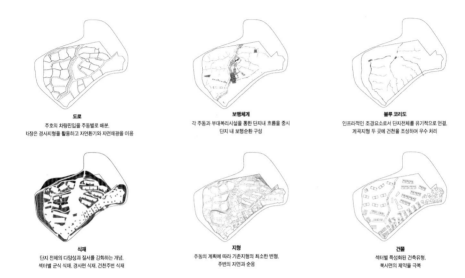

도로
주호의 차량진입을 주동별로 배분.
차장은 경사지형을 활용하고 자연환기와 자연채광을 이용

보행체계
각 주동과 부대복리시설을 통한 단지내 흐름을 중시
단지 내 보행순환 구성

블루 코리도
인프라적인 조경요소로서 단지전체를 유기적으로 연결.
계곡지형 두 곳에 건천을 조성하여 우수 처리

식재
단지 전체의 다양성과 질서를 강화하는 개념.
섹터별 군식 식재, 경사면 식재, 건천주변 식재

지형
주동의 계획에 따라 기존지형의 최소한 변형.
주변의 자연과 순응

건물
섹터별 특성화된 건축유형.
북사면의 제약을 극복

地文的图层。
Layers of landscript.

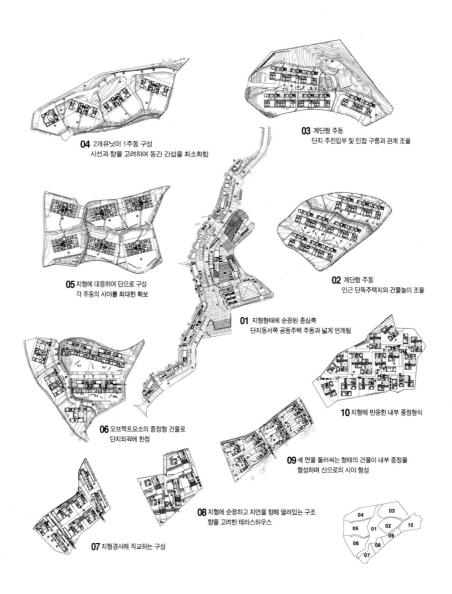

04 2개유닛이 1주동 구성
시선과 향을 고려하여 동간 간섭을 최소화함

03 계단형 주동
단지 주진입부 및 인접 구릉과 관계 조율

05 지형에 대응하여 단으로 구성
각 주동의 시야를 최대한 확보

02 계단형 주동
인근 단독주택지와 건물높이 조율

01 지형형태에 순응된 중심축
단지동서쪽 공동주택 주동과 넓게 연계됨

10 지형에 반응한 내부 중정형식

06 오브젝트요소의 중정형 건물로
단지외곽에 한정

09 세 면을 둘러싸는 형태의 건물이 내부 중정을
형성하며 산으로의 시야 형성

08 지형에 순응하고 자연을 향해 열려있는 구조
향을 고려한 테라스하우스

07 지형경사에 직교하는 구성

分地块设计。

Designs by sectors.

沿着过去的道路设计公共设施。

Communal facility designed through old road.

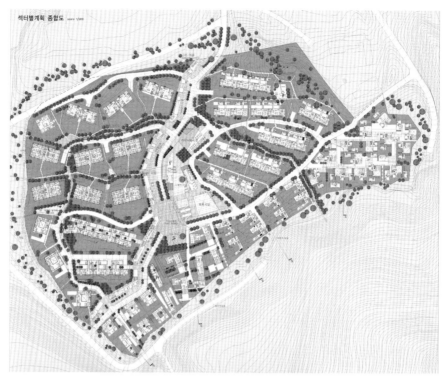

规划方案（上）和现状照片（下）。

Proposal (top) and existing condition (avove).

北京前门大街历史保护区整治规划，2007 年
与弗朗西斯·萨宁共同合作

2007 年的一天，北京的一个开发商找我商量这个项目的时候，我多少有些兴奋。历史上北京最重要的地段之一——前门大街的整治更新项目，想必对任何一位建筑师来说都是非常有吸引力的项目，尽管如此，我还是很想强调我是对这种类型项目最感兴趣的建筑师。

北京故宫的中轴线，不仅是北京城市的主轴线，又是中国人的精神轴，是国家的象征轴。位于这中轴线上的前门大街，古时是皇帝出城赴天坛祭祀时经过的御路，因此叫天街，目前此中轴线向北延伸至故宫北侧的奥林匹克公园。前门大街周边地区的历史可以说是与北京的历史同步的。因位于故宫

Redevelopment of Qianmen Dajie Historic District, Beijing, 2007
In Collaboration with Francisco Sanin

One day in 2007, I was very excited when an intimate client from Beijing came to discuss this project with me. The fact that the project was to redevelop Qianmen Dajie (前门大街), historically one of the most important locations in Beijing, should have been appealing to any architect, but I could not stress enough how much I was interested in such project.

The main axis of the Forbidden City in Beijing is not only the city's backbone but is also a spiritual pillar for the Chinese and a symbolic one for the nation as a whole. Qianmen Dajie that lies on top of the axis—now extending to the Olympic Park located north of the Forbidden City—was called the Emperor's Path as it was a road that the Emperor passed to give religious services at the Temple of Heaven. The area of Qianmen Dajie has been there throughout Beijing's history. As the

前门大街。

Qianmen Dajie.

MASTERPLAN

地文。

Landscript.

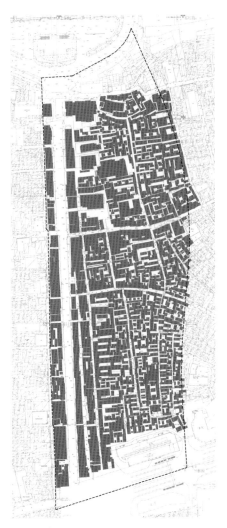

原有的城市组织。
Existing urban fabric.

新的城市组织。
New urban fabric.

入口处，此地区成为商业中心区。在古时候，外地来的官员为了解决进京应试举子的住宿问题，根据自己本地的传统风俗习惯建立符合自己生活习惯的住处，所以这里除了北京本地的建筑形式之外，也留有各地会馆的建筑形式。

这次整治更新项目范围包括长800米的前门大街和其东侧23万多平方米的地区。这里至今还保留着传统民居——四合院，但大部分是民国时期以后变样的居住结构，还有一部分在近代化过程中被破坏掉，被代替的是西洋式的建筑。

过去几年，北京的城市风貌有天翻地覆的变化，在城市的开发过程中许多北京传统特色的胡同也渐渐消失，但此地有充分的理由必须保留传统风貌，因此在制定着非常严格的城市规划。

我在访问现场的第一天就几乎把所有的建筑都调研了一次。原住民已被迁移，胡同内到处都是像废墟一样的空空的房屋，我不禁惊叹。民国时期以

main entrance and exit of the Forbidden City, it has been the center of commerce and was a place where local government officials or students from outside the capital formed a living while preserving their traditions. Its buildings represented the uniqueness of Beijing's architecture but also maintained its own regional characteristics.

Recently, a total of 23 hectares to the east including the 800-meter-long Qianmen Dajie was designated as a redevelopment area. Although some of the traditional houses, Siheyuan（四合院），were still standing, most of the houses had modified structures built after the Republican Era, while others were western style buildings built in the process of modernization.

During the development process, in which Beijing witnessed a change of their city landscape inside and out for the past several years, the street of Hutong（胡同），a traditional residential area, has been torn apart. However, as this area is geopolitically significant, there were more than enough reasons to preserve the area, and a very strict city plan regulation was in effect.

On my first day of visit, I looked at almost all the houses. As the residents had already moved

规划模型和细节。

Model view and the details.

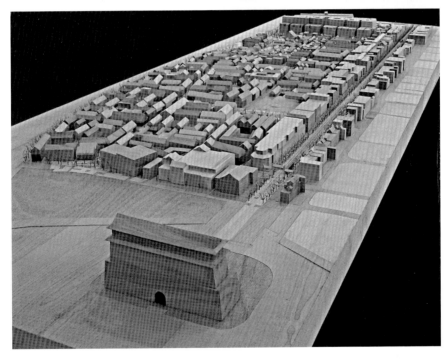

精品酒店规划区。

Boutique hotel block.

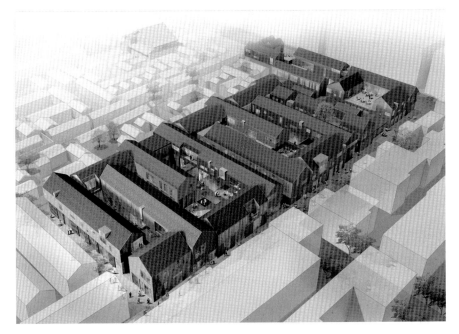

条码型商业区规划。

Bar-code type shopping block.

后，在一套传统的四合院里住着多户人家，四合院的空间结构是非常有趣并多样化的，每个空间都各自有自己的特殊性。空间长或窄，深或短，连绵不断或突有挡墙，几乎一整天的调研我都在细听这些空间"诉说"的故事并沉溺于街巷中。这里的空间是宝库，是生活的智慧。是的，它本身就是一部纪录片，是部连续剧，确实有很高的保留价值。

我在现场对开发商说"地文"的概念，并强烈表达了要保留这些外部空间的意愿。外部空间的保留意味着要维持原有的建筑尺度，这对在此地区策划好新业态——商业设施的开发商来说是一件很困惑的事情，即便如此，灵敏的开发商想起自己小时候经历的美丽的空间记忆，欣然接受了我的建议。她就是长城脚下的公社项目的开发商张欣 (Zhang Xin) 女士。

我先着手画出现状图，果然现状图呈现出非常精巧的空间组织。大部分街巷宽度虽然都不到 2 米，但因弯曲的道路形态滋生出非常丰富的公共空间。

out, the houses were empty and almost in ruins, but I could not hide my astonishment while walking through the area. The spatial structure of the traditional Siheyuan was extremely interesting and diverse as it had been greatly changed by the many households that had lived in it since the Republican Era. Every space was meaningful and unique. Some had a long shape, others were narrow. Some were deep or short, and some seemed to be connected forever while others suddenly came to a dead-end. For almost half a day, I wandered about the streets and indulged myself to the stories told by these spaces. The place was a treasure of spaces and it also gave insight to the wisdom of life. As the place was at once an authentic documentary and a well-structured drama, there was every reason to preserve it.

I immediately stressed to the client that the external space must be preserved, citing the example of "landscript." The suggestion to preserve the external space, which in turn meant that the size of the existing buildings also had to be maintained, may have been absurd for a developer planning to build a new shopping mall, but the intelligent client who remembered the beauty of space when growing up in a rural area accepted my idea. The client was Ms. Zhang Xin, the developer of SOHO China

可是北京市政府已经制定的整治规划中道路宽度为 6 米和 8 米，这会破坏和切断原来的空间组织，但已无法挽回，并且新规划的功能业态所需要的最小室内空间尺寸也是不能忽视的重要因素。

我们首先分析了整体的组织，划分出 4 种不同的城市空间类型（urban patterns）：迷宫型（Labyrinth），主要是不断变化的四合院空间组织，有着最丰富的形态；条码型（Bar-code），其形态和尺度可以满足新的业态功能；保留比较完整的传统民居形态，可以恢复为传统民居，我们将其命名为步行区（Precinct）；前门大街两侧已经形成的建筑聚集地区，我们叫作城市体块（Urban Mass）。

将根据前期分析画出来的新总图（黑色图 / 肌理图）和原来的现状图重合起来，就能出现很多白色部分，从这里可以看出已经最大限度地保留了过去的空间组织。可能要消失的红色部分的原有空间组织，也可以在地面铺装

who was also responsible for the influential Commune by the Great Wall project.

I first drew the current outline of the area, which presented a truly sophisticated spatial structure. Most of the roads were not even two meters wide but it formed various types of curved public spaces. Although the requirement of six-meter- and eight-meter-wide roads prescribed in the city government's redevelopment plan cut off and damaged the existing structure, we had no choice but to accept it. We also had to abide by the requirement of minimum indoor space dimensions of the program.

The overall structure was categorized into four urban patterns: Labyrinth, Bar-code, Precinct, and Urban Mass. Labyrinth referred to the spatial structure of Siheyuan, which had gone through multiple transformations and was thus most diversified. Bar-code concerned volumes where a new program was applied to the current form, while Precinct was applied to areas with relatively sound form of traditional houses, which could be restored as traditional residential areas. Finally, urban mass referred to the already concentrated area along the side of Qianmen Dajie.

上印刻原来的空间肌理，这样至少能够保留记忆碎片。

这种新总图其实在白纸状态下是无法画出来的，我是按照现有的空间组织，根据新的功能业态适当进行变换而画出来的，是在一篇历史悠久的文字之上增添几句我自己的想法获得的结果，只靠我一人是无法创作出如此丰富的空间的。

还有另外 3 位建筑师也被邀请参与此方案设计，首先要最先启动的前门大街的两侧建筑，要求再现民国时期的建筑面貌，所以在这部分的设计中我被除外，我负责的是整个地区其他大部分地块的方案设计。后来我看到北京奥运会的马拉松比赛经过前门大街，此处为中国的国家象征轴，所以也会理所当然地设计这样的马拉松路线。

为了这个项目，我参照了诸多关于保留历史街区方面的文章。

The newly drawn blueprint (the drawing in black), based on the factors mentioned above, was similar to the original, and when placed on top of the original drawing, many areas were left white, meaning that most of the existing spatial structure was preserved as much as possible. A marking was to be made on the land for the existing structures that were to be torn down (shown in red), in order to at least leave a trail of its memory.

The blueprint could not have been drawn from scratch. It is a slightly changed version from the existing structure to adopt new programs. It was like adding a touch of my thoughts to a long-standing story. As a result, spatial abundance, which could not have been created by the effort of an individual, was formed.

Three other architects were also invited for the execution plan. The construction along the side of Qianmen Dajie, which needed to be completed in a short period, had to reproduce the buildings of the Republican Era. Therefore, rather than participating in this primary phase, I took charge of implementing the execution plan on most of the block of the overall area. The marathon course

《关于建筑遗产的欧洲宪章》（1975 年 10 月欧洲议会）

　　1. 欧洲的建筑遗产不只是我们认为最重要的纪念物，而是包含我们城市中的小建筑群和自然或人工环境中的特色村落。

　　2. 在建筑遗产中被印刻的过去，提供给我们的是均衡和完整的生活中不可或缺的环境。

　　3. 建筑遗产是不可替代的，有着精神、文化、社会、经济方面等价值的资本。

　　4. 历史地区会帮助协调社会均衡性。

　　5. 建筑遗产对教育也起到了很重要的作用。

during the recent Beijing Olympics passed Qianmen Dajie, a somewhat obvious plan as it is a symbolic national pillar.

Various documents regarding the preservation of historic villages were referred to in the course of this project.

European Charter of the Architectural Heritage (Council of Europe, October 1975)

1. The European architectural heritage consists not only of our most important monuments: it also includes the groups of lesser buildings in our old towns and characteristic villages in their natural or manmade settings.

2. The past as embodied in the architectural heritage provides the sort of environment indispensable to a balanced and complete life.

3. The architectural heritage is a capital of irreplaceable spiritual, cultural, social and economic value.

4. The structure of historic centers and sites is conducive to a harmonious social balance.

《对历史地区的稳定和现代作用的建议》（1976年内罗毕联合国教科文组织总会）

    ……历史地区不管在哪里都是我们日常环境的一部分，但作为形成此环境的过去的留存体，需要在社会多样性的生活背景中谋求变化，才能获得价值，进而使我们达到另一个不同的层面。

    ……历史地区传递着世世代代文化性、宗教性、社会性活动最分明的证据，因此将历史地区安全地保留并与现代社会相结合，这是城市规划和地区开发中的基本要素。

    ……目前时代到处有着复制和没个性的设计，这种过去时代的鲜活证据，在人类和国民的生活方式的表现和身份认同的确立上，有着宿命性的重要性。

5. The architectural heritage has an important part to play in education.

Recommendation concerning the Safeguarding and Contemporary Role of Historic Areas (UNESCO General Conference, Nairobi, 1976)

…Historic areas are part of the daily environment of human beings everywhere… represent the living presence of the past which formed them… provide the variety in life's background needed to match the diversity of society, and… by so doing they gain in value and acquire an additional human dimension.

…Historic areas afford down the ages the most tangible evidence of the wealth and diversity of cultural, religious and social activities… Their safeguarding and their integration into the life of contemporary society is a basic factor in town-planning and land development.

…In face of the dangers of stereotyping and depersonalization, this living evidence of days gone by is of vital importance for humanity and for nations who find in it both the expression of their way of life and one of the corner-stones of their identity.

《历史小镇的保护》（1987 年华盛顿 ICOMOS 总会）

● 原则和目标

保护的内容包含村庄或者城镇地区的历史性质，以及表现这种性质的物质和精神载体，特别是包含如下内容：

a) 因宅基地和道路形成的城市肌理。

b) 与建筑、绿地、公共空间的关系。

c) 建筑外形和尺度、样式、施工、材料、颜色、装饰等建筑里外的全部内容。

d) 村庄或者城镇地区和邻近自然或人工环境之间的关系。

e) 村庄或者城镇地区通过长期时间积累的多种功能。

Charter for the Conservation of Historic Towns and Urban Areas (Washington Charter, ICOMOS Conference, 1987)

• Principles and Objectives

Qualities to be preserved include the historic character of the town or urban area and all those material and spiritual elements that express this character, especially:

a) Urban patterns as defined by lots and streets.

b) Relationships between buildings and green and open spaces.

c) The formal appearance, interior and exterior, of buildings as defined by scale, size, style, construction, materials, color and decoration.

d) The relationship between the town or urban area and its surrounding setting, both natural and man-made.

e) The various functions that the town or urban area has acquired over time.

● 方法和手段

5. ……保护规划要决定保护哪些建筑，特别是要考虑在意外条件下哪些部分会有消失的可能性。

8. 新的功能和活动要与村庄或城镇地区的性格有竞争性。……

10. 新建建筑或者改造原有建筑时，要尊重原有的空间排列，特别是尺度关系很重要。要在与周边环境的协调中引入现代因素，如果新引入的要素能在丰富这原有地区方面有贡献，就不能禁止此举动。……

12. 在历史地区，交通需要控制，停车场不能损坏原有历史空间组织和环境。

13. 城市规划、地区规划要建设主干道时，不能贯穿历史保护地区。……

• Methods and Instruments

5. …The conservation plan should determine which buildings must be preserved… and which, under quite exceptional circumstances, might be expendable.

8. New functions and activities should be compatible with the character of the historic town or urban area. …

10. When it is necessary to construct new buildings or adapt existing ones, the existing spatial layout should be respected, especially in terms of scale and lot size. The introduction of contemporary elements in harmony with the surroundings should not be discouraged since such features can contribute to the enrichment of an area. …

12. Traffic inside a historic town or urban area must be controlled and parking areas must be planned so that they do not damage the historic fabric or its environment.

13. When urban or regional planning provides for the construction of major motorways, they must not penetrate a historic town or urban area.…

威廉·莫里斯 (William Morris) 的 SPAB 宣言文，1877 年

如果它（历史建筑）不符合现代功能时……与其改变或者扩张历史遗迹，不如建造其他新建筑。……这样我们才能从束缚我们的想法中解脱出来，还能够保护我们的建筑遗产，进而将它们恭敬地传授给我们的后代，让后代也吸取教训。

The Manifesto of SPAB (Society for the Protection of Ancient Buildings) by William Morris, 1877

If [a historic building] has become inconvenient for its present use… raise another building rather than alter or enlarge the old one…. Thus, and thus only, shall we escape the reproach of our learning being turned into a snare to us; thus, and thus only can we protect our ancient buildings, and hand them down instructive and venerable to those that come after us.

悠久的未来： 杨柳青历史区再生计划

    城市尺度的杨柳青大运河国家文化公园项目，是在 2020 年 5 月天津市西青区政府委托 CBC 进行的国际大师邀请赛中，由我们履露斋与瑞安造景、Urban Transformer 策划团队联合进行的方案获胜。

    杨柳青位于天津市西部，于 7 世纪的隋朝开始建设，明朝时期基本建设完成的杭州到北京段大运河共计 2700 千米，杨柳青段包括南运河和北运河，杨柳青因地势低洼多湿地而柳树居多，故得名杨柳青。

    杨柳青大运河国家文化公园规划设计包括元宝岛以及元宝岛北侧的至今传统房屋保留尚多的杨柳青历史文化古镇，元宝岛南侧的文化小镇三部分。

An Ancient Future: Yangliuqing Historical Area Regeneration Plan

The Yangliuqing Grand Canal National Cultural Park is an urban-scale project, for which the government of Xiqing District, Tianjin commissioned CBC to hold a global invitational competition in May 2020. IROJE architects and planners formed a team together with Seo-ahn Total Landscape and Urban Transformer for a proposal that won the first prize.

Yangliuqing is an area located in the west end of Tianjin. The area was developed in the Sui Dynasty in the 7th century and was completed during the Ming Dynasty. Along the Grand Canal from Hangzhou to Beijing, covering a total length of 2,700 kilometres, Yangliuqing is located at an intersection between the South Canal and the North Canal. Yangliuqing is directly translated to 'green willows', attributing to the willows growing in low wetlands.

现状。

Existing surrounding.

场地的历史记忆。

Historical memory on site.

元宝岛段大运河故道在靠近杨柳青古镇一侧，后因裁弯取直形成新的现元宝岛南侧的河道，后来又重新开挖了故道，形成了元宝岛四周环运河的现状。元宝岛、杨柳青历史文化古镇以及文化小镇三部分的占地面积分别为 88 公顷、38 公顷、62 公顷，共计 188 公顷。

　　元宝岛内目前只有一栋文昌阁古建筑，但此处曾经有书院等文化设施和密集的居住设施等，是较为繁盛之处。挖通元宝岛南侧运河前，此处多为湿地与水池，有着自然与建筑融为一体的风景记忆。北侧的历史文化古镇，由过去的传统建筑和街巷形成了牢固的肌理组织，但有些撑不起历史的重担，需要新的活力；元宝岛南侧的文化小镇原来是农耕用地，需要开发利用。

　　历经相同的时代，但镌刻不同纹理的这 3 个地块，需要不同的解决方法，因此我们团队提出用"杨柳青历史区再生计划"(Yangliuqing

The project consists of three parts: Yuanbao Island, which became an island as a result of the construction of a straight canal connecting the two southern points of the meandering canal, which was the center of logistics due to its many curves, the historic reserve where many traditional houses are still concentrated on the northern side of Yuanbao Island, and the cultural town on its south. The three covers an area of 88ha, 38ha and 62ha respectively, totaling 188ha.

Currently there is only one ancient pavilion in Yuanbao Island – the humble Wenchang Pavilion, but this used to be an area with flourishing cultural facilities such as academies along with dense residential buildings, making it a prosperous place. Prior to an excavation of the canal on the south side of Yuanbao Island, there were mostly wetlands and ponds here, providing the locals with a memory of an integrated landscape of nature and architecture. The historic reserve on the north maintains a solid urban structure with traditional buildings and old streets, but it seemed to be struggling with the weight of history and there was a recognition that a new vitality was in need. The cultural town on the south also used to be agricultural land, requiring some development for utility.

Historical Area Regeneration Plan, Y-HARP) 的名字代替了"杨柳青大运河国家文化公园"，用名字浓缩了设计特性。

为了再生 (regeneration) 悠久的土地，根据每块地的特性树立3种策略。

虽然元宝岛内过去村落的形态已全都消失，看似白纸状态，但此处有着消失的建筑和街巷的记忆，要重组 (reorganise) 这种痕迹 (old fabric)，比在此处建立无依据的建筑单体更加紧要并有原真性。这就要通过承孝相的"地文"(landscript) 的建筑语言来实现。

北侧的历史文化名镇需要注入活力，为了通过再活化 (re-habilitation) 达到原有传统村落的再生，将小设施插放在村子里，就像东方医学中的针灸术不同于外科手术一样，将小画廊、微型演出厅、小公园等设施放置在村落中相当于人的脉一样的空间里，由此整体能够找回活力苏醒过来。这就是"城市针灸术"(urban acupuncture)。

Having lived through the same era but developed in different ways thus engraved with different textures, the three areas required separate solutions that will bring about new life. Therefore, our team proposed a different name for the project, Yangliuqing Historical Area Regeneration Plan (Y-HARP), to contain and highlight the characteristics of the design, instead of the original name (Yangliuqing Grand Canal National Cultural Park).

For regeneration of lands with an ancient history, we based our design on three established strategies according to the attributes of each area. Although the old villages of Yuanbao Island no longer exists and seems to be in a tabula rasa state, there were memories of vanished buildings and streets. It was more essential and authentic to work through a re-fabrication of the old fabric than to construct groundless new building. We were aspiring to realise this through the architectural language of 'landscript' by Seung H-Sang.

The historic reserve on the north needed some vitality. Proposed solution was a re-habilitation, where small facilities are installed or injected to the town for regeneration of traditional villages. Just

元宝岛南侧的文化小镇是新开发的场地，这里是以剧场或美术馆等大型文化设施为中心的整体开发 (re-structuration)，要再生此处，需要的并不是强调这些中心设施，而是要打造协调的整体景观，需要文化风景 (culturescape) 的概念。

　　地文、城市针灸术、文化风景是这个项目的核心词和概念，将这些使用在所有部分，做出总体规划。元宝岛是这个项目的重点，也是要先启动的地块。在大师邀请赛中获胜后，西青区政府要先启动元宝岛区域，我们接着做第二阶段的总体规划，于 2020 年 7 月提交了成果。

　　元宝岛的规划的关键是恢复过去的地形和空间结构，我们通过对过去的地图和文献的深入挖掘来掌握地文，重构过去空间结构，再重现湿地和水池，以宜人的尺度形成街巷和区域。

like an acupunctural treatment for the veins of the town as opposed to a surgical operation, we took the approach to bring in small galleries, micro performance halls, small parks and other facilities to be placed for revivification. This is so called the 'Urban Acupuncture'.

The cultural center on the south side of Yuanbao Island is a newly developed site. It aims to achieve a 'Culturescape,' a plan for a coordinated landscape, rather than an outstanding central facility. It is a concept of regeneration through an overall re-structuration of centralized large-scale cultural facilities such as opera theatres and art galleries.

Landscript, Urban Acupuncture, and Culturescape are the key words and vocabulary of the concept which shall be applied to every scale in order to complete the masterplan. Yuanbao Island is the focus of the project in particular, and the area was first target to be achieved. After the competition, the government of Xiqing District wanted to develop the Yuanbao Island first, therefore the masterplan phase 2 was progressed for Yuanbao Island, and the result was reported to the government in July 2020.

以柳口路为界，元宝岛分为西岛和东岛。西岛有国际花园大赛用地，多为预留地，即便预留区域亦根据过去村落的空间和道路划分。西岛除了花园大赛及其配套设施外，还有位于湿地水边的露天和室内音乐厅。室内音乐厅兼有温室的功能，提升了个性（identity）。

东岛有杨柳青传统文化——年画博物馆和相关的文化设施和配套设施，这些单体并非大体量建筑，而是具有村落肌理的多个小尺度建筑的组合，整体打造小村落的风貌，其空间结构同样根据地文形成。

元宝岛整体是步行专用区，因此在车流量较大的柳口路上，局部加盖做成公园，由此来连接东岛和西岛。又通过重塑地形，文昌阁将浮在水边，崇文书院将滨水而置，与丘陵状隆起的跨桥公园一起打造有特色的村落轮廓。

The key to the planning of Yuanbao Island is to restore the original landscape and spatial structure. Through in-depth analysis of the maps and documents of the past, we tried to explore the landscript and reconstruct the spatial structure of the past, hence reproduced the wetlands and ponds, and form streets and areas with a pleasant scale.

Traversed by Liukou Road, Yuanbao Island is divided into the west side and the east. The west side was given to accommodate the international garden expo in the future, hence most of which are reserved for flower garden plots and they are arranged according to the old paths of the village in. In addition to the garden expo and its supporting facilities, the West part is proposed with open-air theater and indoor concert hall along the canal. The indoor concert hall is including greenhouse and improves the identity.

The east side features the traditional culture in Yangliuqing. There are the New Year Picture Museum and other cultural facilities. These single buildings are not large-scale complex, but a combination of multiple small-scale ones suitable for the village-scale context, which their spatial

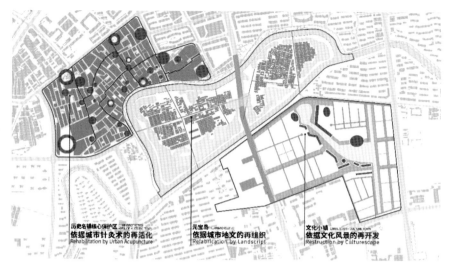

历史名镇核心保护区 二级保护地区 THE PROTECTED AREA IN OLD TOWN
依据城市针灸术的再活化
Rehabilitation by Urban Acupuncture

元宝岛 YUANBAO ISLAND
依据城市地文的再组织
Refabrication by Landscript

文化小镇 CAMAL SCOT-201 THE TOWN
依据文化风景的再开发
Restruction by Culturescape

整体概念构思。

Configuration of concept.

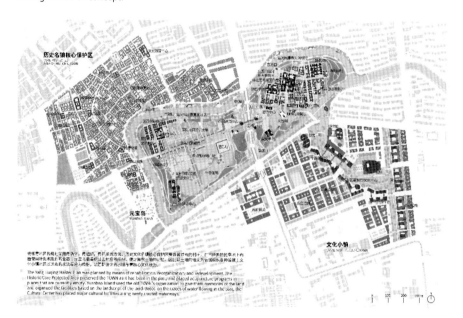

历史名镇核心保护区
THE PROTECTED AREA IN OLD TOWN

元宝岛
YUANBAO ISLAND

文化小镇

杨浦青少区历史规划采用再开发、再组织、再开发的方法。历史文化名镇核心保护区保留了过去的乡土特村。广场目前是空的。元宝岛利用老镇的土地景观来组织设施，使之保留有土地的记忆。围绕以土地历史方式创造的水色种道度；文化小镇利用这些水系布置主创设施，围绕着设计为新建水系布置主要核心文化设施。

The Yang Jiuqing Maldive Plan was planned by means of rehabilitation, reorganization, and redevelopment. The Historic Core Protected Area preserved the TOWN as it has been in the past and placed acupuncture programs in spaces that are currently empty. Yuanbao Island used the old TOWN's organization to give them memories of the land. onic organized the facilities based on the landscript of the old TOWN and based on the traces of water flowing in the past, the Cultural Center has placed major cultural facilities along newly created waterways.

设施布置规划。

Site plan.

东岛鸟瞰图。
Bird's-eye view of East Area.

鸟瞰图。
Bird's-eye view.

元宝岛有 3 处车行入口。西岛西侧和东岛东南侧的车行桥直接与地下车库连接，车库贯穿东西岛，可通往地面各区域。因柳口路现状车行较多，结合后期地铁出入口仅设置应急停车场出入口。

景观在这个项目上非常重要。从 20 世纪末开始，元宝岛上的建设导致原有地形的破坏，此次元宝岛的规划要恢复过去的地形，还原固有的自然环境。因此我们恢复水池和湿地，并用由此挖来的土方打造东西两侧的山坡和裨补林，最终打造新杨柳青十景，体现悠久的未来之风景。

这种项目不只一处，在很多地方发生过类似的事情，包括未来也会有。我希望此项目的规划能成为回顾这些过程的范本。这片土地承载我们所有

structure is also based on the landscript.

Yuanbao Island is mainly a pedestrian area, so on Liukou road with heavy traffic flow, a park is planned so as to connect the east and west side. By reshaping the topography, Wenchang Pavilion will float on the water, and Chongwen Academy will be located by the waterfront to create a distinctive village outline, together with the hilly park.

Yuanbao Island has three accessing points for vehicles. The vehicle accessing via bridges on the west of the west side and on the southeast side of the east side are directly connected to the underground. The garage runs through the east and west side and leads to various areas on the ground. Due to the current situation of Liukou Road with heavy traffic flow, emergency entrance and exit of the parking lot are placed according to the position of the entrance/exit of the subway to be developed in the later stage.

Landscape is very important in this project. Since the end of the 20th century, construction on Yuanbao Island has led to the destruction of the original topography. The proposed plan for Yuanbao

人的生活，所有的城市和建筑也建立在土地之上，因此土地是我们生活的根本，同时也是传给后人的悠长的叙事史。

Island should restore the topography and the inherent natural environment. Therefore, we will restore the wetlands and ponds, and use the excavated earth to build hills and ecological forests on the east and west sides. In result, ten feature landscapes have been proposed, turning into an ancient future.

Similar issues have happened in many different places, and will happen in the future. I hope that the planning of this project will serve as a model for reviewing these process of urbanism. Lands support the lives of all of us, and all cities and buildings are built on lands. Lands are not only the foundation of our life, but a long narrative history passed on to future generations.